The Mindful Universe

OTHER TITLES IN THIS SERIES:

The Art of Mindful Baking

The Art of Mindful Birdwatching

The Art of Mindful Gardening

The Art of Mindful Reading

The Art of Mindful Silence

The Art of Mindful Singing

The Art of Mindful Walking

Einstein and the Art of Mindful Cycling

Galileo and the Art of Ageing Mindfully

Happiness and How it Happens

The Heart of Mindful Relationships

The Joy of Mindful Writing

The Mindful Art of Wild Swimming

The Mindful Man

Mindful Crafting

Mindful Pregnancy and Birth

Mindful Travelling

Mindfulness and Compassion

Mindfulness and Music

Mindfulness and Surfing

Mindfulness and the Art of Drawing

Mindfulness and the Art of Managing Anger

Mindfulness and the Art of Urban Living

Mindfulness and the Journey of Bereavement

Mindfulness and the Natural World

Mindfulness at Work

Mindfulness for Black Dogs and Blue Days

Mindfulness for Students

Mindfulness for Unravelling Anxiety

Mindfulness in Sound

The Mindfulness in Knitting

The Practice of Mindful Yoga

Zen and the Path of Mindful Parenting

"What you seek is seeking you"

"As you start to walk on the way, the way appears"

Rumi

The Mindful Universe

A journey through the inner and outer cosmos

Mark Westmoquette

ant me the
erenity
accept
things
annot change,
courage
change the
ngs I can,
the wisdom
know the
fference"

"Love with all your heart"

Seek and you shall find

"Beyond a wholesome discipline be gentle with yourself. You are a child of the universe no less than the trees + the stars you have a right to be here"

hese pains you feel are messages listen to them"

"Go placidly amid the noise + the haste, + remember what peace there may be in Silence."

Max Ehrman 1927

Leaping Hare Press

First published in the UK and North America in 2020 by

Leaping Hare Press

An imprint of The Quarto Group

The Old Brewery, 6 Blundell Street, London N7 9BH, United Kingdom

T (0)20 7700 6700

www.QuartoKnows.com

British Library Cataloguing-in-Publication Data

A catalogue record for this book is available from the British Library

ISBN: 978-0-7112-5283-7

This book was conceived, designed and produced by

Leaping Hare Press

58 West Street, Brighton BN1 2RA, United Kingdom

Publisher DAVID BREUER

Art Director JAMES LAWRENCE

Editorial Director TOM KITCH

Commissioning Editor MONICA PERDONI

Project Editor ELIZABETH CLINTON

Design Manager ANNA STEVENS

Designer GINNY ZEAL

Illustrator MELVYN EVANS

Printed in China

1 3 5 7 9 10 8 6 4 2

Contents

Introduction

Do you ever get the feeling that something in life just isn't right? Maybe it's too hectic, painful or dull. Whatever it is, something seems to be missing, and we feel a subtle, simmering dissatisfaction in the background. In this book I will take you on a journey to the furthest reaches of the Universe and back, on a quest to quell this dissatisfaction. What will we discover? Well, no prizes for guessing that our most important discovery will be mindfulness. I invite you to explore two questions with me: thinking about our inner selves, what might that reveal about the Universe? Thinking about the Universe at large, what might we discover about ourselves?

STARTING OUT

Because you have this book in your hand I'm going to assume that you are interested in finding out what mindfulness is and how it feels to live mindfully, and that you are also curious about space, astronomy and our place in the Universe.

EACH CHAPTER WILL LOOK INTO a different area of mindfulness to see how that can help us experience the world and the cosmos in a new way. I'll be touching upon ideas from physiology, psychology, spirituality and physics, but you don't need any particular knowledge of these subjects, nor do I assume that you've done any mindfulness practice before. Of course, if you have then so much the better.

Over the course of this book, I will encourage you to think about such things as: awareness; distractions; pain and suffering; how our mindsets and perceptions affect our experience; the reality of time and space; and what it is that distinguishes you from the rest of the Universe. However, this book is not about explaining concepts or theories, it's about developing a first-hand, experiential knowledge of what it is to 'be' in this vast Universe of ours. To that end, I've included a number of guided mindfulness practices for you to follow. I strongly encourage you to do them, not just read them. I will invite you to examine a leaf, which will enable you – as you think about the sunlight it photosynthesizes – to contemplate the

incredible nuclear reactions that go on inside the Sun and all the stars that energize the Universe. I'll ask you to spend some time sensing your body, exploring what it might mean to feel 'grounded', and I hope to add some wonder to your experiments with facts such as: the elements carbon and nitrogen, that make up our bodies, were formed inside distant stars billions of years ago. I will invite you to experience the present moment, to think about what distracts us from it, and what time really is. By doing this, you will see again and again how perceptions, expectations, limited senses and fixed mindsets colour our understanding of reality. Ultimately, we'll be looking into the question of who we really are and seeing how that question can lead us towards a whole new way of perceiving the Universe.

> *The study of astronomy and cosmology . . . is deeply connected to our humanity*

My own journey started with the study of the outer Universe. Then, for various reasons that I'll explain, I gradually turned my attention to the equally vast inner universe. It's my wish that, by the time you finish this book, our exploration of the connection between the two universes will have shown you that the study of astronomy and cosmology isn't a purely intellectual pursuit of interest only to people in ivory towers. It is deeply connected to our humanity, purpose and sense of curiosity about the world

around us. Ultimately, I hope this journey will show you ways to live day-to-day, here on planet Earth, with greater contentment, compassion and wisdom.

A Note About Mindfulness

Mindfulness means bringing 100 per cent awareness to what's happening in this moment. It also means not judging what you find to be good or bad, nice or ugly, better or worse, or wishing it were different. Paying attention is relatively easy, and very important. If you get distracted when chopping a carrot then sooner or later you're going to need a plaster, perhaps even stitches. Not wanting things to be different, however, is much harder. Even if you're paying attention to the carrot you may find yourself wishing the carrot were bigger, the knife sharper, or that you weren't chopping a carrot at all. Wishing and wanting lead us out of the moment.

Just because we have the ability to be mindful, it doesn't mean we do it very often, or do it voluntarily. Can you remember a time when you were trying to concentrate on something but your mind kept on thinking about something else? Maybe you were trying to write a report but were replaying the memory of a difficult meeting; or maybe you were listening to someone talk but actually thinking about whether they could see that coffee stain on your shirt. When this happens we're not being mindful, we're distracted and likely being judgemental. What about times when we're on

'autopilot'? For example, getting to work and not remembering whether you locked the front door; or losing your phone and later finding it in the fridge. Being on autopilot is the opposite of being mindful. That's not to say it's a negative thing (without it we couldn't drive a car whilst holding a conversation) but problems arise when we live too much of our life on autopilot.

Without knowing about mindfulness, many people seek out activities that force them to be in the present moment, to be so completely absorbed in something that their normal distractions, worries and criticisms don't arise. Without absorption, an untrained mind is easily and involuntarily swept off into a fantasy land of thoughts, ideas, plans and memories. Thinking isn't bad in itself, but when we're in a whirlwind of thought it's all too easy to get sucked into a cycle of negativity, catastrophizing and self-criticism. That's when we start feeling stressed and depressed. Learning mindfulness helps us to focus on the present moment, anywhere and anytime, without the need for absorbing activities. Importantly, it also teaches us how to be okay with that present, no matter how tricky or uncomfortable it might be. Then we can act with awareness, rather than on impulse. These two aspects – awareness and acceptance – are the key to finding lasting happiness and contentment.

Mindfulness as a practice is present in many traditions and religions, even if that word itself isn't used. However, since Buddhism placed it at its centre from the very beginning, it is

most closely associated with Buddhist practice, and it is within Buddhist traditions that it has become highly refined. In the original language of Buddhism the word *sati* is used, which literally means 'remembering' or 'recollection', and relatively recently – in the mid-twentieth century – the English word 'mindfulness' was used to translate it. Being mindful means remembering to notice what's happening, here in the present moment, and remembering that every moment is extraordinary – it will only happen once in the whole of eternity.

My Personal Journey into Mindfulness

I can't say I ever consciously sought out mindfulness. It found me. The practice seeped in slowly, like walking in the mist and gradually realizing that you're totally soaked. Mindfulness has catalyzed some of the biggest changes in my life, regarding both my inner perspectives and outer orientations, and has changed the whole course of my life.

When I was very small I was sexually abused by my father. My mum found out when I was six and he was arrested. A few years later, when I was 9, my mum remarried. Then, when I was 13, my mum and stepdad were in a car accident that killed my stepdad and put my mum into hospital with first-degree burns over most of her body. To save her life they amputated both her legs and one arm.

I've come to understand that kids who encounter these painful experiences can react in various ways: they can, for example, go off the rails and use drugs or dangerous behaviour to numb themselves, deny reality or assert some level of control upon their lives. They can become hypersensitive, co-dependent or disruptive, or they might shut themselves off emotionally, becoming cold and distant. I did the latter and found myself at university, aged 20, rather lost and struggling to make friendships. I had decided to study astrophysics. In my first year of study I started going to a yoga class at the gym because I'd been told it would help with my target rifle-shooting (which I'd got into in a big way at the time). I struggled to wrap my scientific brain around yogic concepts such as *prana* (energy/*chi*) but the practice was enough of a physical challenge to keep me returning.

At 25, I'd never had a romantic relationship, and relations with my family were formal at best and, at times, hurtful. My mum recommended that I start seeing a psychotherapist. I found someone through the university counselling service and he was great; we met twice a week for more than three years. It was at this time that I also met my Zen teacher, who was also a yoga teacher, having just taken over teaching the class I was going to at the gym. He ran a meditation group and invited along anyone who was interested to go. My years of yoga had sparked a curiosity about meditation – I'd noticed a sense of ease and contentment often arose during *savasana*

(the final relaxation period in a yoga class) and I craved more of it. I started going to the Zen classes and, soon after, made a commitment to a daily meditation practice which continues to this day.

Over the next couple of years, through the mindful lenses of Zen, yoga and psychotherapy, I started to discover all those emotions I'd been shutting away. I saw how my unconscious had learned to hold people at arm's length in order to protect myself from getting hurt. I found the repressed anger I had towards my abusive father and the off-duty policeman who caused the car accident. I realized how desperately I wanted my mum to be the way everyone else's mum was, and came to see how part of my healing was to accept that she never would be and to grieve that loss. I took a yoga teacher training course in 2009, and on this course I managed to let my walls down sufficiently to laugh more wholeheartedly than I ever had before.

At its root, mindfulness is a technique for letting things be and letting things go

Mindfulness was the central thread that ran through all my practices. It was the tool that allowed me to let all my emotions and realizations arise, be seen and acknowledged, and in that acknowledgement I knew I didn't have to 'do' anything with them. At its root, mindfulness is a technique for letting things be and letting things go. We can't change events

that happened in the past; all we can do is see them and accept them for what they are and, by doing so, become increasingly content with the way things are right now.

When my wife and I met, I'd been doing research in astrophysics for more than a decade. Over the next few years I began to notice a subtle current of dissatisfaction arising around my path in academia. I realized I'd become addicted to the results of research, the highs that come from publishing those results and the fear-driven push for more, for better, for bigger that is common to many strands of modern life. The academic way of life had caused me to fall out of love with astronomy. I realized that I wanted to put my relationships first and do something more immediately helpful and people-focussed. So, in 2013, I switched tracks and took up teaching yoga and mindfulness full-time.

Over the years I've come face-to-face with both the outer reaches of the vast cosmos and the inner depths of my being. More recently, I've also rediscovered my love of astronomy. In the pages that follow, I'd like to share with you some of my discoveries so far, and offer a few signposts that will point you towards a more mindful Universe.

CHAPTER ONE

Paying Attention: to the Sun & Beyond

In response to the trauma of my childhood, I put a great deal of attention into my academic studies – but I see now this was my way of distracting myself from a painful reality. Mindfulness gives us a way of developing and improving attention as a means of seeing life more fully, rather than focussing on one thing at the expense of another. Attention can become an amazing tool for exploring the wondrous reality in front of us, noticing things we've never seen, in new and exciting ways.

Paying Attention

I remember being told as a child to 'pay attention' when my mind wandered off: I don't remember anyone ever offering to teach me how.

To a certain degree, we do know how to pay attention – we develop an instinctive ability to direct our mind and use our will to hold our attention on something. But more often than not (especially if what we're doing isn't that enthralling or, perhaps, is uncomfortable or difficult) our mind wanders off, again and again. We end up thinking about things that make us happy, things we'd rather be doing, or start worrying about things we've done or have yet to do.

I've often wondered, given my childhood traumas and the imprints they left on me, how I did so well academically. After all, other people with similar backgrounds often don't. Looking back now, I see that I used my academic work in astronomy as a distraction. Over time, my brain worked out that, if I buried myself in my studies, I could avoid having to face other, more painful aspects of my life. At the same time, it made me feel good because I was achieving something. The unconscious behavioural adaptions I developed in response to life's ups and downs were positive with regards to my academic work, but over time negatively affected large swathes of the rest of my life. I ended up 'heady', emotionally distant, and with skewed priorities with regard to family and relationships.

Avoiding Pain

As humans, we'll go to extreme lengths to avoid or numb ourselves to pain – particularly emotional pain, which may include psychological aspects that might threaten the identity we have created for ourselves. Our protection mechanisms are strong and can often operate under the radar of our conscious mind. We can get ourselves into the most extraordinary pickles as a result of a series of unconscious reactions to trauma, and to bonds of attachment, that we might have experienced in early life – none of which were our fault, or even of our doing.

One of the effects of desperately wanting to avoid pain is that it makes you concentrate (hard) on something else. Because my unconscious mind decided that academic work was the best distraction, I developed good skills in that particular kind of concentration. For another person it could have been surfing, playing the violin or shopping (none of these activities are inherently bad, of course, but if we use them to distract ourselves from what we're really feeling then they become maladaptive behaviours). I say 'that particular kind' of concentration because I can remember, in some psychotherapy sessions, how I was very poor at concentrating on a memory or finding my way to the feeling that I had about the past, or about the way I had acted earlier that week – because it was painful. My brain would want to distract me from that onto something else, something abstract or 'heady'.

Thankfully, childhood trauma is not a prerequisite for developing good concentration, nor is distraction from pain the only method for improving it. Mindfulness teaches us a balanced and non-harmful way of developing and improving this foundational skill for life.

Concentration

Getting better at concentrating has knock-on benefits. I can't think of any pursuit that wouldn't benefit from better concentration – whether it's writing an email, chopping an onion or driving a car. The thing is, though, not only were we not taught any practical ways of developing better concentration at school, but some aspects of modern life are designed to actually disrupt our concentration. Adverts, for example, are specifically made to grab your attention away from whatever you're doing. In a large city like London, for example, research has shown that in one 45-minute journey, a traveller will be exposed to more than 130 adverts, featuring more than 80 different products. Phones are cleverly designed to tug continuously on your attention with their social media notifications and messaging pings. And the situation seems to be getting worse. In response to people's shortened attention skills, the length of newspaper paragraphs and TV news bulletins are getting shorter. For example, on American TV the average length of a political soundbite has dropped, from 43 seconds in 1968 to 9 seconds in 1988, to under 8 seconds

in 2011. Films, documentaries and TV programmes move faster than ever before (you'll notice this if you watch a film as little as 20 years old – it feels slow and ponderous compared to contemporary films). Even a one-minute video on social media seems far too long to watch in its entirety these days! We read quickly, try to multitask and get totally scatter-brained as a consequence. A study carried out by Eyal Ophir et al in 2009 showed that habitual media multitaskers are "more susceptible to interference from irrelevant environmental stimuli" (i.e. more easily distracted) than occasional multitaskers. Modern society and culture is encouraging us to develop short, fragmented and diluted attention spans.

Working with my own distractions is still a daily task. As I write this, I can feel my attention being pulled towards what I am going to have for lunch, to my phone to check social media, to the email ping I just heard – and to the worry that I won't have enough time today to get everything done. We're all affected. We might believe that multitasking is a good thing, but research by Joshua Rubinstein et al in 2001 showed that nobody can effectively do more than one complicated thing at a time.

Mindfulness practice increases your concentration skills. You become distracted, your mind wanders off, you bring it back, and repeat. Slowly, our 'muscle' of attention gets stronger, just as our bodies do going to the gym. Not only is mindfulness practice an antidote to the forces that eat away at our

MINDFULNESS EXERCISE

MINDFULNESS OF THE BREATH

This short 'mindfulness of the breath' meditation is based around developing concentration.

- Take a moment to notice how you're breathing right now. There's no need to change it or influence it in any way – just notice the natural movement in your body.
- You might notice the flow of air; mostly in your nose, in your mouth or in your throat. You might notice your chest rising and sinking, or you might feel the movement mostly in your belly. There's no right or wrong. Take a minute or two to focus on how it feels.
- If at any point you get distracted and start thinking of something else, know that this is fine. It's normal. Just as soon as you notice, redirect your attention back to your breath.
- Breathing in and breathing out.
- Is your breath lengthening or shortening? Is it speeding up or slowing down? Is the in-breath shorter than the out-breath, or is it the other way around? There's no better or worse, just notice your experience as it is.
- And any time you get distracted, just gently let go of whatever it was you were thinking about and come back to the breath.
- Can you experience your breath (and your breathing body) with curiosity and openness, simply noticing your in-breath and out-breath?

Take as long as you want with this.

ability to concentrate; it is, in itself, the cultivation of the art of concentration that can unlock in us the potential to take it ever further, ever deeper.

Wandering Off

Mind-wandering isn't bad. Sometimes, for instance, when we're trying to problem-solve or be creative, the free association of mind-wandering is just what we need. But the point is that, when the mind wanders off involuntarily, we lose touch with the present moment. More often than not, rather than being creative, we end up grappling with anxious thoughts or lost in made-up fantasies, which almost never helps. Mindfulness is about getting back in touch with the amazingness, the raw-ness and the vibrancy of reality, right here.

The practice 'Mindfulness of the Breath' invites you to concentrate on the feeling of your breath. As you do this, you will become more aware of its subtleties. Perhaps you notice how deep the breath is, which parts of the body move, and how the rhythm changes. You might feel awkward when you first try noticing the breath without actually controlling it. Perhaps you notice a tendency to want to control it, or a sense that it 'should' be this way or that way. Perhaps you get distracted time and again and have to really work to bring your attention back.

Whatever you experience, it is your experience. There's no right or wrong, should or shouldn't. Being aware of how

you are, and accepting it, isn't that easy. Yet, as you train your attention muscle, you also train your neighbouring twin muscles of non-judgement and acceptance. You might say, 'I was so distracted by the fly buzzing around in the room I barely noticed three breaths in that whole five minutes' – fine, no worries. That's the way it was. You were doing your best and that was your experience. The trick is to be 100 per cent okay with it. When you can be 100 per cent okay with the fly buzzing around and distracting you, then you are on the way to being okay with uncomfortable or painful feelings, emotions or memories when they arise.

When you pay attention, you see that attention itself is an amazing function of the mind

When you pay attention, you see that attention itself is an amazing function of the mind. With barely any effort you can choose to become aware of things inside the body, such as the breath, the weight of your body, feelings of hunger, and many other sensations. Equally, you can become aware of things outside the body. For example, look at a wall near you right now. Notice its size and colour, the shades, textures and shadows. Notice if there's anything obscuring it; notice what your attention is drawn to. This is being mindful of an external object. Moving the attention around like this is natural and easy. Holding it there and remaining interested in a wall for a period of time, however, is quite another matter!

Following a Leaf to the Sun

After observing something with a sharp, focussed attention, which is good for seeing detail, we can slowly broaden and soften our focus so that, gradually, we see more and more. If we let it, that broad attention can take us to the furthest reaches of the Universe.

Our attention can be tuned in a similar way to a torch, where twisting its end narrows or widens the beam. A narrow, sharp focus highlights details and specifics, whereas a broad, diffuse focus allows relationships and context to come into view. Both types are important for increasing your awareness of what's going on in this moment. Practising using your attention in these two ways is like going to the gym and working the physical body in different ways. For example, first we might run on the treadmill, then we might lift weights. Mindfulness is like a work-out for your mental 'body'.

Taking the example of a leaf, a narrow focus allows you to see the detail of the colours, textures and shapes on its surface. A wide focus allows you to start seeing it in its larger context. As we broaden and soften our focus, we can start to see more dimensions of this leaf. We see the broken-down minerals and nutrients collected by the plant's roots. The roots are embedded in the surrounding soil, which is made from rock eroded by the wind and rain, mixed with dead and broken-down plant matter. So, in the leaf, there's the soil

MINDFULNESS EXERCISE

SHARP EYES & SOFT EYES ON A LEAF

For this practice you'll need to find a leaf and some kind of timer. The leaf might be on a tree in your local park or on a plant in your garden. Try not to just visualize a leaf – find a real, living leaf that's close enough for you to see it in detail. For the timer, you might want to use the stopwatch on your phone. A good friend of mine calls this practice 'sharp eyes and soft eyes'.

To begin, settle your body in a comfortable posture, making sure you can see your leaf. The closer the better. Set your timer for three minutes.

- In this first period we're going to examine the leaf with sharp eyes. Look at it in detail – explore the colours, the textures, the shapes, the light and shadow, the veins, any deformities or damage, and any other details you can see. Really go into it in depth. If your mind wanders off, as soon as you notice this simply come back to your leaf. Don't waste time judging yourself; just carry on with the practice.
- When the timer is up, relax and let your eyes rest for a moment or two.
- Now, reset your timer for another three minutes and, this time, look at your leaf with soft eyes. If your focus was previously like a laser beam, now it's going to be more like a soft, diffuse beam. Keep your eyes on the leaf, but start to broaden your view and see it in its context. How is it situated? How does it look next to the other leaves? Can you see the rest of the plant and, now, the wider surroundings in your peripheral vision? Try blurring your eyes slightly – how does the leaf look now? What made you chose this particular leaf? Does it remind you of anything?
- When the timer is up, relax again and give yourself a moment to notice how that felt.
- What's the difference between looking with sharp eyes compared with soft eyes? Did you notice any difference in your posture between the two? Which one took more effort? Which one held your interest most? There's no right or wrong, just your experience. Everyone finds something different.

and the matter of countless worms and bacteria that have lived and died to create it. The veins also transport water up from the soil. Perhaps only hours ago that water fell as rain from a passing cloud. And how did the cloud come about? The planet's water cycle of evaporation and condensation is driven by heat received from the Sun, via the circling of air through the atmosphere and sea currents through the oceans. So, with a broad open focus, we can 'see' the planet's vast oceans and swirling clouds right there in the leaf's surface.

Photosynthesis

Perhaps the most wondrous of processes we can see right there in the leaf is that of photosynthesis. Photosynthesis is a process of 'transduction'– the conversion of one type of energy into another. Back in the dim and distant past, plant organisms worked out that they could convert the abundant carbon dioxide in the atmosphere into chemicals that it could use to grow, combining it with water using the energy of light. Photosynthesis happens in the green chlorophyll proteins in the cell of a leaf. As a by-product of the process, oxygen is created and released as waste.

That plants have evolved to make direct use of sunlight energy is amazing. The journey that this light (and all the light irradiating the Earth's surface) has been on since its creation inside the Sun is also incredible. Just one average leaf receives around 10^{18} (10 with 18 zeros after it) visible light photons

from the Sun every second. But only about 5 per cent of these photons (packets of light energy) are used by the leaf for photosynthesis; the rest are either absorbed as heat or reflected. When you look at your leaf, it's these reflected photons that you see.

Every one of these photons will have had to pass down through our atmosphere in order to make it to the leaf. If it had been an ultra-violet or an X-ray photon it would have been absorbed by the atmosphere (luckily for us, as these are deadly) but our visible light photon made it successfully to the ground, past all those aeroplanes and clouds. It's very likely to have been scattered, here and there, by dust particles or water droplets – particularly if it was originally a blue light photon (blue light is scattered more easily than red, which is why the sky is blue). Even before that – before entering our atmosphere and before passing all the orbiting space debris and satellites – the photon would have traversed the 150 million kilometres between us and the Sun in about eight minutes, travelling at the speed of light.

The photon traversed the 150 million kilometres between us and the Sun in about eight minutes

So, turning back to your leaf: the photon that enters your eye and enables you to see it – this was, quite literally and just minutes ago, zipping through interplanetary space from the Sun, past Mercury and Venus, to the Earth.

Deep Inside the Sun

To understand where this photon came from, we need to understand a little of the Sun's structure and energy processes. In the deep core of our Sun, under immense pressures and temperatures, enormous numbers of hydrogen nuclei (single protons stripped of their accompanying neutron) fuse to create helium, each releasing energy. This is known as the 'hydrogen fusion reaction'. The energy released initially takes the form of extremely high-energy gamma ray photons.

Outside the Sun's core is a layer called the radiative zone. As these high-energy photons enter this layer, the density of particles is so high that they can't travel more than a few millimetres before they are absorbed by an atom and immediately reemitted. This happens countless times and the outward progress of the photons is slowed to a veritable crawl (even though they are still travelling at the speed of light between each encounter). The radiative zone is 300,000 kilometres deep and is so dense that photons take upwards of a few million years before they escape to the next layer. In this absorption-reemission process, photons collide, causing them to lose much of their energy. The vast bulk of them slide from being high-energy gamma rays to medium-energy visible light photons. The next layer of the Sun, the convective zone, is still 200,000 kilometres thick but it's far less dense, so a photon takes only about ten days to journey through it. Once it is through, it reaches the photosphere (the surface).

So, the photons that are entering your eye, having raced across the inner solar system, traversed our atmosphere and reflected off your leaf, have actually spent millions of years before that, bouncing like pin-balls through the layers of dense plasma in the Sun. That green light from the leaf intimately connects us to the inner workings and conditions inside our Sun, backwards through time.

We're not just connected: we are this Universe. Nothing is separate. No one and no thing.

The mass of the Sun is roughly 10^{30} kilograms, three-quarters of which is hydrogen, with the rest mostly helium. Even though it converts 4 million tons of matter into energy every second through hydrogen fusion, it's been doing this for more than 4 billion years so far. And, believe it or not, the Sun is a fairly sedate, comparatively small star, which means it shines fairly dimly and lives a long time. Its fuel reserves and energy consumption are such that it will continue to shine just as it is for another 5 billion years.

Feeling Deeply Connected

You might think that I was filled with daily awe and bliss when, as an astronomer, I was studying such things in detail, and that my mind was constantly blown by the enormity and mystery of the Universe. Sadly, I found the awe and mystery

faded quite quickly in the daily grind of research (only to reoccur on special occasions, such as giving public talks or discovering something exciting in my data). Like most of my colleagues, I learned to study and observe in a rather dissociated way. Obviously, such an attitude has its uses – it allows one to reduce, analyse and separate things out – but when it becomes the main mode of thinking, it can persist into other areas of life, where it can be less helpful, especially when it is allowed to continue unchecked.

I've found no better antidote than regular mindfulness practice to counter this tendency. A practice like 'sharp eyes, soft eyes' not only reawakens my sense of magic in the present moment, it also reminds me of the scale and reach of the Universe, that is always near enough for me to touch.

All we need to do is pause in our busy, distracted, hectic lives, and to spend a few moments directing our attention towards something as common as a leaf and, if we can concentrate for a minute or two, we can see it all – the soil, the rain and weather, sunlight shining through space and the vast ball of energy from which it came. We are deeply connected to everything in this vast Universe of ours. In fact, we're not just connected: we *are* this Universe. Nothing is separate. No one and no thing.

CHAPTER TWO

FEET ON THE EARTH

The nineteenth-century Scottish physician and African explorer David Livingstone was once attacked by a lion. The extreme stress of the experience, he wrote, made him feel like a 'mouse after the first shake of the cat'. For most of us, the stress we encounter isn't as extreme, but it can be equally dangerous. Without sufficient time to rest, the effects of even minor amounts of stress can accumulate to the point where we feel hot-headed, disconnected and ungrounded. Mindfulness helps us slow down, de-stress and sense the ancient and precious ball of rock beneath our feet that we call home.

Busy, Busy, Busy

When we're busy and stressed, we're at our most ungrounded. We feel flighty, rushed, preoccupied, out of touch and disconnected from our bodily feelings (particularly the more subtle ones). Sadly, though, it's a place many of us inhabit more often than we'd like.

Here's a scenario: you've had a hectic day at work and there are a million thoughts flying around in your head. You've got half an hour to eat before you're off out again to meet some friends. As you grab items from the fridge and wait for the oven to heat up, your mind is playing through the emails that have been bouncing to and fro all day. You ought to call your sister but you just won't have time this evening. There's a vague feeling of tension in your shoulders but the sensations are only on the very edges of your awareness and some part of your brain is hoping they'll just wear off.

When we feel overwhelmed by the demands on our body and mind, we say we're 'stressed'. But why do we get stressed?

Stress & Threat

Our bodies have evolved over millions of years to respond to potential external threats with a rapid mobilization reaction. This involves a plethora of instinctual responses that have saved humanity's bacon countless times over the millennia. However, modern life doesn't bring us face-to-face with big,

one-off threats (like wolves, snakes or sudden precipices) very often; instead it pummels us with many, many 'micro-stresses'. Our bodies have the same reaction to the 'threat' of bad traffic holding us up on the way to an appointment as they do to the war cries of an enemy clan (albeit – hopefully – to a slightly lesser degree).

The problems really arise when you compound that bad traffic 'threat' with all the others that arise in a normal day: your phone running out of battery; loud building works outside; no milk in the fridge; having to cover for someone off sick; returning from your tea break to 45 unread emails; the boss in a bad mood, and so on. All these minor stresses accumulate because there's not enough time between each incident for your energy and hormone levels to return to baseline. By the time you get home, you're so wound up it might take you the rest of the evening (and perhaps some of the night) to wind down enough to sleep.

When the brain and nervous system perceive a challenging incident as a threat, you enter survival mode. It's important to know the body can react in different ways depending on the perceived level of danger: if the challenge is considered low-to-medium risk, then you experience the rapid mobilization reaction, otherwise known as the 'fight/flight' response. The adrenal glands release various hormones into the blood stream, including adrenaline and cortisol. These hormones boost heart rate, blood pressure and muscle tension. In turn,

the hormone glucagon is released (as insulin is supressed) so that stored energy in the form of glucose becomes available. You're on high alert and ready for action – to fight the potential threat or run away. To most people, the energy of this state tends to feel buoyant but it also causes that top-heavy, hot-headed, ungrounded feeling. The body becomes tense as it resists gravity and gets you ready for confrontation. The mind races; anxious, catastrophizing thoughts multiply, and you feel 'all up in the head'. All this surplus energy is looking for a way to be released, often causing you to react impulsively and not always considerately.

Neuroception

If the brain perceives the level of threat to be totally overwhelming, then the brain-body may go into 'shutdown'. The term 'neuroception' was coined by one of the pioneers in the field of neuroscience, Dr Stephen Porges, to describe our threat assessment sense. In his book *The Polyvagal Theory*, he emphasizes that the activity of neuroception is wholly subconscious, but conditioned by past conscious and unconscious experiences and beliefs. Consider the following scenario: Whiskers the cat has caught a mouse and, rather smugly, brought it to show you. The mouse is hanging limp in its mouth and appears to be dead. As it was caught, the mouse's involuntary response was to enter shutdown mode. Whiskers deposits it on the floor and, because it's not moving,

loses interest momentarily and looks up at you as if to say, 'What do you think?'. The mouse notices Whiskers glancing away, and in that moment reanimates and runs away through the gap under the skirting board. Going into shutdown has saved the mouse's life – as it has for humans countless times over the course of history. Humans, like mice, don't have big claws, razor-edged teeth or protective armour. Like the mouse, if attacked by an overwhelmingly stronger person or persons, the human neuroceptive circuits perceive there's no way to fight and win, and so initiate shutdown.

Some of the most obvious signs of shutdown in humans include a collapsed body posture, weak, cold limbs, or a numb 'spaced out' feeling. The wish to curl up in a ball (which children often do) is another sign. It's exactly what happens during stage fright. Other signs might include lowered heart rate and blood pressure, feeling dizzy or nauseous, needing to defecate or urinate (in extremis, doing so uncontrollably), feeling constriction around the throat and having difficulty getting words out. We might also have a sense of feeling trapped or disconnected from the world; we might lose some of our body awareness, have a muddled or fuzzy mind and have difficulty remembering what happened later. So, when under apparently extreme threat, going limp helps to mitigate any serious injury that might be sustained if we were to resist; similarly, it is thought that when the brain closes down, it is protecting us from potentially lasting emotional pain.

I saw the lion just in the act of springing upon me [...] Growling horribly close to my ear, he shook me as a terrier dog does a rat. The shock produced a stupor similar to that which seems to be felt by a mouse after the first shake of the cat. It caused a sort of dreaminess, in which there was no sense of pain nor feeling of terror, though quite conscious of all that was happening.

DAVID LIVINGSTONE (1813–1873)
PHYSICIAN AND CHRISTIAN MISSIONARY (FROM HIS LOGBOOK, 1849)

Have We Been Educated Out of Our Bodies?

Something seems to have happened to us as modern humans that echoes our instinctual reactions to perceived threat: we have become increasingly disconnected from our bodies. For a variety of reasons, we've become numb to the gamut of complex and subtle signals it produces. Western-style education is one guilty party. Someone once remarked to me that average state education seems to concentrate further and further up the body as we go through childhood. Education in the first years at school is more-or-less whole-bodied, but by the time you get to university the only thing that matters is the brain. We are practically educated out of our bodies.

That was certainly the case for me. I always enjoyed maths and science, chose those subjects at college and sealed my fate by opting to study astrophysics at university. My science-based education taught me that reason and logic (generally perceived to be activities confined to the brain) are ideals to strive for, while intuition and emotion (which have significant body-centred components) should not merely be put aside but actively discouraged. Science is conducted with cold reason alone, is it not? Every step is closely considered and carefully thought through with no room, apparently, for awe or inspiration. It took me a long time to realize that, indeed, this dismissal was only apparent. It was just my perception – the surface layer covering a much deeper and unspoken truth.

When applying for jobs I was expected to show passion for the subject, and there was an underlying recognition that I'd become a scientist because I was, at some level, in awe of nature and seeking to understand its hidden depths. My peers and I all knew about the famous examples of scientists in the past coming to major new understandings through distinctly non-rational means: Archimedes jumping out of the bath, shouting 'eureka' and discovering the law of buoyancy; Isaac Newton's apple falling on his head, prompting his understanding of gravity; August Kekulé's dream preceding his discovery of the structure of benzene; and Albert Einstein's imaginary journey on a beam of light, prior to his formulation of the Special Theory of Relativity. But I don't remember intuition being mentioned once when we were learning the skills of being a scientist.

I THINK THEREFORE I AM?

It's important to appreciate the dichotomy between intellectual thought (talking about something) versus embodiment (talking as something). This perceptual split reflects back to the philosophical system of Descartes who solidified the perceived separation of body and mind when he claimed cogito ergo sum, 'I think therefore I am'.

INTELLECTUAL THOUGHT IS THE JOB of the brain, whereas embodiment involves the whole, integrated mind-body system. Spurred on by Descartes' claim, Western society in the eighteenth century increasingly came to see the body as primitive, base and carnal: the source of urges that lead to sin, emotions that lead to irrational behaviour and intuitive knowledge that was unreliable and irreproducible. In contrast, the intellectual mind was seen as the seat of reason and logic and therefore superior and pure. Though we now know that this philosophy is faulty – contemporary research shows that aspects of emotion and feeling are indispensable for rationality – but it soaked sufficiently into the cultural psyche over the centuries to continue to influence our thinking today.

At university I felt like a brain on legs. My body was there to bring nutrients to (and sometimes to remove alcohol from) my brain, and to transport it from my bed to the lecture room, to the office and back. I sat for hours in front of my computer, so engrossed in work that I often missed rising hunger sensations or pain in my hand from so much typing. Little did I know that years of feeling stressed and ungrounded and being out of touch with my body were promoting long-term ill-health. Had I done my research I would have seen that unconscious stress and a lack of body awareness has cumulative negative effects on a whole range of physical factors, from sensory acuity and blood pressure to the quality of our chromosomes, fatigue and, therefore, likelihood of burnout, cardiovascular disease and a host of other diseases.

A society in which people live with high and persistent levels of stress and muscular tension, that places a huge emphasis on thoughts and the thought-based world, is an ungrounded society. Research shows that, when we slow down, connect to others and change our attitudes and habits around stress, we see improvements in our physiology: better cardiovascular function, a more balanced hormone system and, even, changes to our gene expression. Although mindfulness isn't the only method known to help with this, it does contain an incredibly effective set of practices.

When we get caught in the whirlwind of head-centred worries and fantasies, just consciously bringing our attention back into our body – and to the feeling of our body touching the ground – can do wonders. We return to our present-moment reality and to the simplicity of body-felt sensations. This is also an excellent way of giving the system a moment to rest and return to its 'base-line' between those daily stresses.

MINDFULNESS EXERCISE

SENSING THE GROUND

• Please take a seat – any seat, wherever you are right now. Sit down in such a way that you feel upright and both your feet can rest comfortably on the floor. Let your body have a sense of poise about it, such that your spine is long and you're not bent forwards, backwards or sideways.

• Lower your gaze and keep your eyes softly open while looking at this page (if you've read this already you can let them close fully). Let any tension in your face gently soften away – eyebrows, cheeks, jaw, lips, chin.

• Become aware of your buttocks and the back of your thighs pressing against the chair. You don't need to do anything, just notice the pressure and weight. If a thought arises about what you're feeling (or about anything else), that's fine – just bring your attention back to the *feeling* of your body-weight on the chair: the pressure, the temperature, the softness or hardness. Notice how the chair is 100 per cent supporting you. Your body weight is transferring down through the chair into the floor. Take a moment. Explore this feeling.

• Now begin to notice the feeling of your feet on the floor. Maybe they're in socks or tights, or inside shoes. Sense whatever feelings you can discern: the pressure, the temperature, any feelings of softness or hardness. Is the weight equal between your left and right foot, or is it more in one foot? Is the weight mostly in your toes or in your heels? Notice the weight of your legs transferring down through your feet to the floor.

• See if you can get a sense of the general state of your whole body. Overall, do you feel light or heavy? Can you feel the gentle tug of gravity? Do you feel tense or relaxed? Is your mind busier or quieter than when you started?

Sending Down Roots of Awareness

From sensing your own body supported by the ground, your awareness can take you downwards into the earth, the ground and outwards to the whole planet. After realizing how connected you are to this ball of rock we call Earth, we can consider how it took its place in our vast cosmos, billions of years ago.

The key to feeling grounded is to remain aware of the feeling of gravity and of the ground beneath us. It's simple, but easy to forget. Sense the downwards transmission of weight through your body. Notice how your feet and the chair legs are connecting to the floor. When I do this, even briefly, I can feel my stresses melting away and my body relaxing. Now continue to reach downwards with your mind: through the floor – perhaps via the floorboards or the structure of the building you're in – down into the ground (if you're reading this on a boat or an aeroplane then try reaching your mind down below the water or air). It may help to imagine roots or tendrils of awareness growing downwards, eventually finding the soft, damp soil beneath. Try to see yourself as not 'on' the ground, but as part of the ground.

This soil has built up over thousands of years, by erosion caused by the wind, rain, rivers and glaciers, and from generations of plants and animals that have decomposed and been compressed together. Keep going down and you'll get to the

bedrock, made from sediment laid down in an ancient ocean or in the fires of a long-extinct volcano. This rock was formed millions, if not hundreds of millions of years ago, and has risen and rolled with the movement of Earth's tectonic plates to its present-day position, right below your feet.

Keep going, sending those tendrils of awareness down, from your body, through the chair and into the ground. What you're sensing and, I hope, feeling ever more part of, is planet Earth. Our beautiful blue-green home; one of eight enormous balls of rock and gas orbiting our Sun. As you've been reading this, the soil and the rock beneath you have been anything but static and immobile. The surface of the Earth is rotating at 1,600 kilometres per hour, causing the Sun to look like it is gliding gently across the sky. Underneath Earth's rocky crust there are layers of semi-liquid and totally molten rock and magma, bubbling and circulating above a solid ferrous core.

Welcome to your ground, your stability, your foundation. It is where your physical body has come from and where it will eventually return. We're not just standing on it, but are part of it. Part of planet Earth. Let yourself be held by it – and let your stresses, tensions, tightnesses and worries melt into it.

The Formation of the Solar System

The Earth, Sun and all the planets in our galaxy formed about 4.5 billion years ago from an enormous cloud of gas – one of a vast number in our galaxy. A perturbation, maybe from a

nearby supernova explosion or collision with another gas cloud, caused our cloud to begin collapsing. As it slowly started shrinking, the slightly denser regions became denser still and the cloud gradually fragmented into clumps. The runaway collapse caused the density and temperature in each clump to rise, and as it shrank it began spinning, eventually forming a flattened disc. Like a centrifuge, the heaviest elements were spun outwards while the lightest elements (hydrogen and helium) remained in the centre. When the temperature in the hydrogen-rich core reached 10 million degrees or so, nuclear fusion began and the cloud ignited from within. To begin with, this nascent star would have been completely hidden in its dense cocoon, shrouded to the Universe outside. Slowly the energy released by the brightening core began heating and blowing away the halo of leftover gas and dust. After a few million years, all that remained of this gas and dust were the densest clumps of, what is now, rock. These fragments crashed into each other; sometimes they smashed each other apart, but sometimes they coalesced like droplets of water, eventually forming spherical balls – that we now call the planets.

Today, 4.5 billion years later, the planets haven't cooled down completely. Volcanoes, earthquakes and tectonic plate movements on Earth are driven by molten rock in the mantle, kept liquid by the energy left over from the planet's formation. Temperatures in the Earth's metallic iron-nickel core

still reach over 4,000°C (7,200°F), which, together with the tremendous pressures, keep the metal alloy in liquid form. It's the convection and rotation currents in the molten iron that cause our planet's magnetic field.

You & the Planet Go Back a Long Way

As distant as all this may seem from our everyday experience, when we become aware of our feet touching the solid ground beneath us it can serve as a reminder that we are made of this ever-changing planet. We are not in any way separate from it. It is strong enough to hold us and support us through thick and thin. It's been around for 4.5 billion years, and alongside all the volcanoes, earthquakes, ice ages and crashing land masses, has cradled the evolution of humanity.

Let's bring to mind the scenario we discussed at the beginning of this chapter: the hectic day at work, with a million thoughts flying around; feeling tense; trying to make dinner and getting ready to go out, all in half an hour. Now imagine if that person took a moment to do a grounding meditation before they left the house. With jacket on, car keys in hand and about to open the front door, they stop. They drop their attention through their head, torso, legs, feet and into the floor. They notice the downwards pull of gravity and spend a moment acknowledging what it's like standing on the ground, feeling held, feeling part of this great, ancient Earth. Imagine the difference that might make to their night out with friends.

If the Earth is 4.5 billion years old then, in a sense, we all are. That's a third of the age of the Universe! To put it in context, if we equate the age of the Universe – 13.5 billion years – to 13.5 metres, every centimetre represents 10 million years. Stepping forward from zero, we'd have to walk 9 metres to get to the point where our solar system formed. As for humanity, we'd have to go a further 4 metres and 49.8 centimetres – that's just 2 millimetres before the end – before evolution formed us, we created language and started talking about feeling grounded. But, without that cloud of interstellar gas that coalesced 4.5 billion years ago, there'd be nothing to talk about and no one doing the talking. We *are* this planet.

Is There Life Out There?

Life is incredibly rare and precious. Each of us is given just one chance at living, and every moment only comes once. Mindfulness teaches us how to savour life, thereby helping us to see the importance of valuing every life we come across.

The earliest undisputed evidence of life on Earth dates from at least 3.5 billion years ago, but it seems very likely that life came about not long after the oceans formed, 4.4 billion years ago. Perhaps it started in a rock pool, only to be extinguished by an earthquake or asteroid strike (common at that time) before starting again in another rock pool. Since then, life has proved tenacious, resilient and prolific on this third rock from the Sun.

So the question arises: if life began relatively quickly on Earth and eventually evolved into beings that could ask questions such as 'Where did we come from?' could that not have happened on other planets?

To calculate the odds of finding intelligent alien life, astronomer Frank Drake came up with an equation in 1961. He said that it depended on seven things: (1) the rate of formation of stars suitable for life (some stars are very unstable or violent); (2) the fraction of those with planetary systems; (3) the number of planets per star with an environment suitable for life; (4) the fraction of those on which life actually appears; (5) the fraction of those on which intelligent life evolves; (6) the fraction of civilizations that develop technology that we could detect; and (7) the length of time those civilizations exist for. Unfortunately, at the moment we have almost no idea how to estimate anything after the first two variables.

What's helping with getting a handle on the second and third points is the discovery of extrasolar planets (or exoplanets – planets around other stars). It's one of the most exciting developments in astronomy in the last 30 years. The first confirmed detection of an exoplanet occurred in 1992 and at the time of writing, there are over 4,000 known exoplanets in our galaxy. The nearest one is Proxima Centauri b, 4.2 light-years from Earth, orbiting Proxima Centauri, the closest star to the Sun.

Life is So Precious

So far it looks like the number of stars in the galaxy with planetary systems is quite high (Drake's equation, point 2). A planet that's suitable for life (3) has an orbit in the so-called 'goldilocks zone' where water may be liquid on its surface. Roughly half the number of exoplanets so far discovered inhabit this zone. But we have no way of knowing the likelihood of life forming, even if water were present. Making educated guesses for the other unknowns, our current estimate gives us about one planet with intelligent life per galaxy at any one time. And we're it! Human life is extraordinarily rare and that makes it very precious. We therefore have a responsibility to do our best to enable it to flourish wherever we can.

Practising mindfulness of the body will help you to discover and acknowledge what being stressed and ungrounded feels like, and will give you more awareness of your own stress reactions and impulsive behaviours. When I'm stressed, practising mindfulness helps me let go of my wish to feel otherwise, and allows me to acknowledge that hot-headed, spacey, unrooted sensation with less judgement. When you have a quiet moment – sitting on the train or walking through a park, perhaps – try letting your awareness drop downward so that you can sense our amazing planet. Contemplating the ground beneath your feet helps you to find perspective and foster a sense of curiosity and gratitude. This is one of the best antidotes to stress, negativity and tension that I know.

CHAPTER THREE

We Are All Stardust

Many people walk around like a disembodied head, on a body that might as well be a foreign land – like those old maps that label the lands beyond what's known: 'Here be dragons'! But it doesn't have to be like this. When we learn to feel our bodily sensations without judging or analyzing them, we enter a whole new way of being. We can make friends with our body, and respond to its messages with kindness. As we gradually redraw the map of our body, we can begin to appreciate that not only are we inseparable from the earth beneath our feet; but that the myriad chemical elements inside us were produced in stars billions of years ago. We are literally made of stardust.

The Sensing Mode

We often only think of our bodies in a negative way; we're critical of the way we look and only pay attention when our body shouts at us (in pain or in pleasure). I remember a time when I was like this too. We can know things in many different ways. Our linear, logical faculty of reason and intellectual thought is good at facts and figures, whereas our intuitive 'gut feelings' and holistic 'right brain' experiences operate in a completely different way.

In the past, my body below the neck felt like an uncharted territory full of feelings that I only had the vaguest notion about. Some things, like hunger, I knew how to deal with, but other things like knots, rumblings, tinglings and constrictions I didn't know how to appease – and I'd really rather they just went away.

The body doesn't have to be a strange world full of potentially disagreeable or uncomfortable things. Mindfulness can show us how to get to know our bodies in a totally non-judgemental way, giving us the space and time to make new friends with our sensations.

A key practice in mindfulness is the body scan, where you get out of your thought and head-based world and into your visceral, body-based world. For example, when I say I 'know' about the feelings in my chest, it's not the same 'knowing' as when I say I know the size of our galaxy. In other languages

> There is no division between the spirituality of the mind and the spirituality of the body; they are both the same, so there is no conflict.
>
> Chogyam Trungpa (1939–1987), Tibetan teacher

different words are used. French and German, for example, distinguishes between *savoir/wissen* (meaning to know abstract or factual knowledge) and *connaître/kennen* (meaning to be familiar with). Sensing my chest has nothing to do with thinking, reasoning or remembering; it's a kind of knowledge that's much closer to intuition.

The Second Brain

When we switch to our sensing mode, we consciously tap into two neurological systems that operate in a very different way to our faculty of reason and intellectual thought. The first is our enteric nervous system (ENS). This is a network of over 100 million neurons (more than in the spinal cord) embedded in the walls of our gut, and has been dubbed our 'second brain' by Michael Gershon in his book of the same name. The primary job of our ENS is to manage digestion (doing so independently of the head brain), but digestion is actually one of the most potentially dangerous things humans do. The ENS has to determine if a particular food is toxic, infected, safe to

MINDFULNESS EXERCISE

BODY SCAN

For this practice, lie down or sit back comfortably. Muscle tension blocks out subtler feelings, so the more relaxed you are the more you'll be able to feel. In this practice simply do your best to sense whatever you find without judging it to be good or bad, or something you want or don't want.

- First, bring your attention to your head. Notice its weight – are you holding it up or are you letting it rest? There's no right or wrong; just notice. What can you feel in your forehead or your cheeks, lips and jaw? If there's any tension or tightness, that's fine – just let it be and notice if it changes. If there are very few, or no sensations, that's also fine.
- Bring your attention to your shoulders. How do they feel? Are they tense, relaxed, up or down? Again, however they are, it's fine. Just notice.
- Become aware of your chest . . . your belly . . . your waist. What do you notice? Take your time – don't just skim-read these words but take a moment to enquire at each area: what can you feel? Sometimes there are strong sensations, sometimes less so, and sometimes the sensations are very subtle.
- Notice what you can feel in your back – your shoulder blades . . . your spine (its shape and curves) . . . your lower back.
- Notice your arms – the bend in your elbows . . . your hands (whatever they're touching – the curl of your fingers, how you're holding this page).
- Become aware of your hips and pelvis. Notice your thighs . . . lower legs . . . ankles, feet and toes. If your feet are in socks or shoes, how does that feel?
- Now see if you can let your attention broaden out to encompass your whole body. Do you feel generally light or heavy? Warm or cold? Tense or relaxed? Remember, there's no right or wrong, no 'should' or 'shouldn't'. As much as you can, let it be just as it is.
- Now take a nice deep breath and have a stretch, if you want. Take a few moments to scribble down what you found on some paper or in your diary. The act of writing can be very useful for making sense of what you feel and for keeping a track of any changes you notice over time.

digest, nutritionally balanced, sufficient/insufficient. It also has to manage our gastrointestinal microbiota (bacteria that live in our gut) that are now understood to play a vital role in our gut health.

Information on the state of the ENS is passed up to the head-brain via the vagus nerve—the tenth cranial nerve, running from your brain stem right down to your abdomen. This information is then interpreted in terms of feelings of safety, satiation, contentment, trust and wellbeing. Feeling butterflies in your stomach, for example, isn't just the *result* of feeling anxious, it's actually one of the important signals we get from our body that tell us about our physical and emotional state – the feeling of butterflies in the stomach is actually due to the redirection of blood away from the gut as part of our fight-flight stress reaction. The body treats the digestion of food as less important than supplying muscles with energy when under threat. The phrase 'gut feeling' isn't misplaced. When the thinking mind is allowed to quieten down and we become aware of our gut feelings, we're connecting our conscious brain to our ENS and listening for the primary sensations that arise from this important area, before cognition or reason get involved. Interoception – as this internal listening process is known – is the precursor to intuition.

We become aware of our gut feelings . . . before cognition or reason get involved

The Master & His Hemisphere

The second neurological system that we have the ability to tap into is described by Iain McGilchrist in *The Master and His Emissary*. This book examines the specialisms that have evolved in the two hemispheres of the brain – the right hemisphere's job, as McGilchrist describes it, is to be inclusive, holistic, and to look at the big picture with an integrative comprehension. In contrast, the left side is all about precision, logic, division and discrimination. When we try to understand something, we would typically start by seeing the broad picture with our right brain. Processes in this hemisphere might pick out something that needs investigating, and so it will hand that information over to the left side, which then applies its detailed, linear processes. The left then hands back the results of its investigations to the right for reintegration and so it can decide what to do next. In modern life, where reason and intellect seem to be held in such in high esteem, there arises some difficulty in that second transaction: once the left side is recruited for its skills in precision and dissection, it doesn't always want to hand its findings back to the right. It wants to have the final word.

In the book, McGilchrist uses the metaphor of a master and his emissary to explain this shortcoming. In a nutshell, the emissary (representing the left hemisphere) develops ideas above his station and, deciding he doesn't need a master (the right hemisphere) anymore, delivers his own messages to

the town that relies upon him. The town descends into confusion because it no longer gets a clear picture of what's going on in the wider world. McGilchrist argues that this is where the West has got itself. Beginning with the ancient Greeks, and accelerated by the Age of Enlightenment in Europe and the modern scientific revolution across the world, we have increasingly emphasized theoretical, precise, categorical thinking at the expense of integrated vision and experience. Despite its inferior grasp of reality, the left hemisphere is increasingly taking precedence in the modern world and is less and less willing to pass its findings back to the right hemisphere for integration. I recognize, now, that that was exactly the situation for me when I was in academic research.

During a mindful body scan, we allow the reasoning, left-hemisphere mind to quieten down. We rest back into our holistic, intuitive mind. We can't be truly attentive to sensations arising in our palms by thinking *about* them; if we can learn to just feel them, we will directly experience them.

Ultimately, the left and right hemispheres of our brain, together with our enteric nervous system and other systems like the brain stem, need to work in synergy, unified and integrated together. But when we find ourselves out of balance, we first have to work on quietening the over-emphasized aspects and cultivating the forgotten faculties. This is what the body scan is all about. As you get more comfortable with sensing your body without thinking about it, you enter the realm of direct experience. With time you'll be able to direct this same kind of attention outwards to your perceptions of the world, thus bringing you the closest you can get to reality (given the limitations and biases of our perceptions that we'll discuss later).

The Body as a Finely-Tuned Antenna

The incessant internal chatter that, more often than not, is our starting point, actually *can* quieten down with mindfulness practice. It is possible to have moments of internal silence where there's space to sense the subtler feelings arising from the body. With more practice, you will become able not only to discern the gamut of sensations caused by the things going on inside, but you will also notice sensations arising in response to or in resonance with things happening outside, from others or from the environment. The body becomes a finely-tuned antenna. You will become more aware of people's expressions, body language, vocal tone and gestures, and you will see the subtleties of their moods, emotions and energetic ups and downs.

When we create the space to listen deeply, we become better humans

When we're not so absorbed by our own habitual melodramas, when we create the space to listen deeply, we become better humans: better lovers, friends, siblings, parents, colleagues, and generally better citizens.

Sensations to Supernovae

Our brain and nervous system – everything we use to make sense of the world around us – are all made of chemical elements that were born inside a star. Stars are the chemical factories of the Universe.

Sensations arise as a result of the brain combining and interpreting incoming electrical signals from the sensory receptors around the body. Let's say you've just touched something hot with your finger. The heat stimulates the temperature sensors in your skin to create an electrical pulse that travels along the connected neurons up to your head. The brain interprets the pulse as 'heat'. If the impulse is strong enough (because the object is very hot) this might cause another electrical pulse to travel back down to the muscles that will pull your finger away from the heat source. These processes involve many chemical reactions including many different elements, charged ions and molecules.

Technically, a nerve impulse is created by an 'action potential'. These are generated when electrically-charged chemicals move through the cell walls of the neuron; the most important chemical elements are sodium (Na^+), potassium (K^+), calcium (Ca^{++}) and chloride (Cl^-). Each one of these elements was once formed inside a star. Our Sun is thought to be a third-generation star that has incorporated all the chemicals made by the stars that came before it. The solar system contains 90 or so different elements that have accumulated here over the aeons from all the explosive forges of the vast Universe. Those elements are all found here on Earth, and some of them make up your own body. Let's have a look at the origins of these elements and particles, and see whether it changes the way we perceive this human body we all have in common. When introspection meets knowledge, amazing things can happen.

A Cosmic Body Scan

When the Universe began, there was nothing but energy. In the first second it cooled sufficiently for the fundamental particles (quarks, electrons, and so on) to solidify, and very soon after that the quarks glued together to form protons and neutrons. Seconds later, protons were able to capture electrons and thus form vast quantities of hydrogen.

In the minutes that followed, the temperature and density was such that nuclear fusion bound together some of these protons and neutrons to form helium (two protons, two neutrons and two electrons) and a small amount of lithium. This process is known as Big Bang Nucleosynthesis.

Bring your attention to your body. Become aware of the watery saliva in your mouth. Move your tongue around, or take a sip of water and notice the feeling of the liquid in your mouth. Water is a molecule with one oxygen atom bonded to two hydrogen atoms (H_2O). The protons forming the nuclei of every atom of hydrogen in your saliva were created within

the first minute or so after the beginning of the Universe. In fact, your body is three-quarters water; for billions of years all the hydrogen in that water existed in space, moving in and out of gas clouds, stars and planets.

After the initial flash of the Big Bang, conditions became too cold for nuclear fusion, and the Universe plunged into darkness for 500 million years or so. The cooling continued until the first hydrogen and helium gas clouds were able to form. In the hottest, densest cores of these clouds, circumstances were right for nuclear fusion to start again. The first stars were born and the Universe lit up once more. It is only in stars that all the other elements were made; first we'll look at how, then I'll reveal the elements of stardust within your body, and inspire you to feel what it's like to be made of them.

The evolution of a star through its lifetime, and which elements it can make, depends on two factors: its mass and its initial composition. The earliest stars would have formed only from the products of Big Bang Nucleosynthesis (hydrogen, helium and a tiny amount of lithium) but, as we'll find out, future generations of stars contained greater and greater concentrations of heavier elements. In astronomy, elements heavier than helium are called 'metals' and the term 'metallicity' refers to the proportion of heavier elements in a star's initial composition. Not much is known about very low metallicity stars since there aren't any around these days. However, for average metallicities, we know that stars with masses less than around eight times that of our Sun (seen as low-mass stars) live a fairly quiet life in the slow lane. Our Sun, for example, is predicted to continue doing what it's doing for another 5 billion years (having lived 5 billion years already). Massive stars, up to 40 times the mass of our Sun, live life in the fast lane, shining bright and hot and burning themselves out within less than 50 million years or so. Extremely massive stars (masses greater than 40 times that of our Sun) are also known to exist. They are fleeting visitors, living for just a few million years, but they can have a tremendous effect on their surroundings due to their extremely high energies.

Death of Stars, Birth of Elements

All stars, big or small, eventually run out of the raw hydrogen that fuels them. The smaller ones fizzle out slowly, while the bigger ones become unstable and explode.

Fizzling stars, in their last gasps, begin fusing helium into heavier elements like lithium, carbon and nitrogen. Sometimes, they throw off their outer layers in slow, pulsating waves, creating transient structures known as 'planetary nebulae'.

Let's try to sense some of this carbon and nitrogen in our bodies. First notice your skin. Perhaps there is heat or coolness in your fingers or toes, an itch, or maybe the weight of one leg as it rests, crossed over the other. The outermost layer of your skin is made of receptor cells (for detecting heat and pressure) and proteins (like keratin), which are made of amino acids, which themselves are made of carbon, hydrogen, nitrogen and oxygen. The bulk of the carbon and nitrogen in the Universe has been synthesized over billions of years in the dying fizzles of small(ish) stars. The carbon and nitrogen making up your skin would have, undoubtedly, once been part of a beautiful planetary nebula of glowing gas enveloping one of these slowly expiring stars, between 5 and 12 billion years ago.

To synthesize much heavier elements, from oxygen up to krypton (including the sodium, potassium, calcium and chlorine found in our neurons), the higher energies of exploding stars known as supernovae are needed. There are two types of supernova explosions. The first occurs at the end of the life of a very massive star – when it finally exhausts its fuel, the energy that has kept it puffed up disappears and the star collapses in on itself. The supernova happens as a kind of rebound: when all this material crashes into the centre it has nowhere to go but outwards again. What's left behind is a 'neutron star' or, if the progenitor was big enough, a black hole. All these epic explosions synthesize, scatter and sow new elements and matter throughout the Universe.

Let's find some of the products of these explosions in your body. Bring your attention to your mouth, become aware of your teeth. Move your tongue around them and sense their shape and hardness. Your teeth are coated in enamel, primarily a crystalline calcium phosphate – the hardest substance in your body. The calcium, oxygen and phosphor constituents were formed from lighter elements that were fused together in the unimaginable heat of past supernova explosions. Most of the oxygen and phosphor in the Universe, including that which makes up the enamel in your teeth, comes from massive star supernovae, whereas calcium is synthesized in massive star and binary star supernovae. Continue to move your tongue around your teeth – it's mind-boggling to think that parts of the surface you're feeling were once inside a supernova!

The second type of supernova happens in a binary star system (two stars in orbit around one another). If one is a smaller star that has come to the end of its life – and is in a phase known as a 'white dwarf' – and the other's orbit is close enough, the white dwarf will begin sucking material from its larger partner. Eventually, the white dwarf can no longer support the weight of the accumulated material, and it collapses, exploding in a supernova.

Now try to sense your heartbeat. You might be able to feel it just as you are without moving, or you might need to put a finger on your wrist or the side of your neck. Notice the pulsing flow; you'll be sensing just a small proportion of the total amount of blood (4.5–5.5 litres/9–12 pints) circulating inside your body. Blood contains many elements, including red blood cells made with a protein called haemoglobin. Haemoglobin contains iron that helps bind oxygen to the cells for transport throughout the body. Iron is surprisingly abundant in the Universe, due to the various ways it can be formed; the

core of our planet Earth, for example, is made of liquid iron. The vast bulk originated in massive star and binary star supernova explosions that will have blasted the newly formed iron back into space – some of which found its way into your blood.

Try also scanning your body for any sensations of digestion. These might be gurgles in your belly or a rumble in your stomach. Digestion is one of your primary metabolic functions. Essential for these metabolic processes are certain proteins and enzymes that are spread throughout your body, all of which use copper. All the copper in your body (as elsewhere on the planet, for example in copper pipes and coins) was formed by the fusing together of various lighter elements in the powerful forges of long-forgotten supernova explosions.

Much of a dying star's newly-made chemicals are expelled into deep space – slowly if the star is fizzling, rapidly if it explodes. These may create new gas clouds that, in turn, collapse, form new stars, live, die and return more newly-made chemicals to space. In this way, the Universe is enriched with heavier elements.

Stars are being born and dying all the time – just one cycling process among the countless going on in the Universe. Humans join in with their own cycles: breathing; circulation of blood; digestion; waking and sleeping; being born, growing old and dying. We play our part in the wheeling, turning and returning story of all things. Just as the Sun will end its life, most probably, as a fizzling star gently blowing its own

newly-created carbon and nitrogen back into the Universe, so at our passing we will give back our elementary particles, that will serve as the foundation for future generations and worlds to come. All endings are beginnings.

Beyond a Supernova

To synthesize even heavier elements – heavier than molybdenum (atomic number 42) – more energy is needed than even a supernova can provide. Step in, neutron star collisions. A neutron star is the remnant of a supernova explosion. With no energy to hold the remaining gas up any more, the immense gravitational forces compress all the leftover electrons and protons to such a degree that they combine into neutrons. Neutron stars are therefore, quite literally, stars composed entirely of neutrons. They are relatively rare, and collisions between them are rarer still, estimated to occur just once every 10,000 years. However, they are some of the most lively events in the Universe, briefly emitting the energy of more than a million Suns. As the neutron-rich matter is ejected, elements from molybdenum up to plutonium (atomic number 94) can form.

Gold (atomic number 79) is one of those elements. Gold is present in your body in very small amounts. Although inert (non-reactive; which is why dentists sometimes use it to fill holes in teeth), some believe it plays an essential role in certain bodily processes. Gold is an exotic material, prized

for its rarity and beauty – but it is especially exotic because it can only be made in the collision of two neutron stars. The presence of gold in your body connects you directly to these inconceivably energetic events. The only way the Earth could have accumulated sufficient gold from these rare and distant events is because the Universe is so immensely old and such a good recycler of material.

Let your attention soften out so that you become aware of the whole of your body – all the parts, processes, reactions, chemicals and elements all together. The elements that you're made of are the same as those found throughout the known Universe. You come from the Universe and will one day return to it. In fact, you never left; and your story, your changes, your cycles are the changes and cycles of the Universe.

Over the years I've found the practice of remembering where my physical body comes from to be invaluable when I find myself caught up in the whirlwind of thoughts, ideas and worries. I drop my attention into my body, tune into my present-moment sensations and do my best not to judge them to be good or bad. Then I recall that every element in my body was once created in the depths of space, millions and billions of years ago. Doing this helps me keep things in perspective.

Through everything you see, touch and eat, and through the very fabric of your human body, you are intimately connected to the vast Universe, its myriad stars and all the billions of years of their living, dying, colliding and exploding.

CHAPTER 4

Wandering Through the Universe

When we become intensely focussed on the present moment, time becomes a curious phenomenon. What is time? Does it really 'flow'? What are the 'past' and the 'future'? Science has shown that time is flexible and the latest theories suggest it may just be an illusion of perception. Certainly, we all perceive it passing at different speeds in different situations, and sometimes it seems to disappear completely. If we learn, more and more, to fully inhabit this present moment, we can find a way of effortlessly experiencing what arises, moment-to-moment. Gradually, we can, as Zen master Eihei Dogen said, 'permeate completely all time'.

THOUGHTS, PUPPIES & RIVERS

When we spend time in silence doing nothing, it's easy to see how quickly the mind grasps around for something to hook onto. It just does it, outside of our volition. You could say the brain is designed to produce thoughts like the heart is designed to pump blood.

PEOPLE HAVE LIKENED AN UNTRAINED MIND to a young puppy. It won't sit still for a moment, doesn't listen to a word you say and likes to get up to mischief. Training a puppy takes effort, patience and consistency – just like your mind.

One of the common misperceptions about mindfulness is that it's about clearing the mind and sitting in tranquil emptiness. This isn't the point at all. In fact, it's actually not possible to stop thoughts – although, with some effort, we might be successful for a moment or two. The traditional analogy is that of a dammed river. The river represents the flow of our thoughts. If we dam the river, yes, the flow stops, but only until the build-up of water behind the dam reaches capacity. At that point, the river breaks over the dam and starts flowing again. There's no way of forcibly stopping the river, just as there's no way of forcibly quietening the mind. The only way to get the mind to calm down is to let it be, and stop agitating it.

Instead of a river, imagine your mind as a jug of water. If we stir up the water until it's spinning in a vortex, no matter how hard we try, we can't get the water to stop spinning again by force. It just swirls around and becomes even more turbulent. But if we stop stirring, it will eventually slow down. All our activity – doing, thinking, planning – acts to stir up the water of our mind. The only way to get it to slow down is to stop feeding its activity. We learn to let thoughts arise but stop engaging with them. Every time you find yourself thinking *about* something, let it go and come back to your point of focus. This is exactly what you've been exploring in the mindfulness practices so far. Have you noticed any moments of relative quiet or stillness? Even fleeting moments of stillness can feel like blissful relief from the busyness of life. But it's important to realize that feelings of bliss, like anything else, will always come and go. They're lovely while they're there, but we let them go, too.

The Mind Is Like Teflon & Velcro

Mind-wandering is following thoughts from one to the next in a long chain. As you sit staring at the computer screen trying to work, suddenly you find yourself fantasizing, reminiscing or day-dreaming. It just happens. But mind-wandering can also be more volitional – you might sit in your favourite easy chair, in the sun, and allow your mind to wander through and around certain topics. Sometimes mind-wandering is useful, like when you're trying to solve a problem, or be creative.

I've also heard it said that the mind is like Teflon to positive thoughts or memories, and like Velcro for negative ones. For instance, you might vividly remember that jibe from your sister when you were 11, but forget what a lovely birthday present she got you last year. If you're edging towards a negative mood anyway, mind-wandering can make things worse. For instance: a memory arises of someone you met yesterday, she had a beautiful necklace that sparkled in the sun; it reminded you of a star twinkling in the night sky; you start hoping tonight will be a clear night; you realize it's been pretty cloudy recently, in fact where you live is a pretty bad place for stargazing, you only moved here because it was convenient for work; work is pretty stressful at the moment, you don't see a way out of it.

Rumination is what happens when mind-wandering gets stuck on a loop. As much as you might want to break out of the cycle, you just can't. It's almost like the mind is revelling in it, repeating that difficult conversation, that insoluble problem or silly thing we did, ad nauseum.

Helpful or not, what mind-wandering and rumination have in common is that they take us away from experiencing the present moment. When we're thinking about something else, our attention is not on what's happening right here and now.

Many things in life pull us away from the here and now. Let's say you're writing a text message to a friend and you hear the sound of a window smashing. Acknowledging the

MINDFULNESS EXERCISE

DEALING WITH DISTRACTIONS

• Arrange your body into a comfortable, upright, sitting posture. Make sure your spine is long, balanced and aligned. Tuck your chin in a little so the back of your neck is long. Let your gaze softly rest on these words. Soften your face and shoulders. Relax into your belly.

• Bring your attention to your breath. There's no need for it to be particularly slow or deep – just breathe normally, naturally. Put your attention on the area where you feel the breath most intensely and try to follow the movement of each breath. Spend some moments doing this before reading on.

• Did you get distracted or start thinking about something else? Did you read on and get distracted by these words? If so, that's absolutely fine. If not, that's also fine too. Return your attention to your natural breathing and continue following the movement. Breathing in, breathing out. If your mind wanders off, just as soon as you notice, gently guide it back to the breath.

• Sometimes the mind doesn't wander much. But other times it can feel like we're being pulled all over the place by reams of thoughts arising, one after another. Neither situation is a measure of how well you're doing. All you need to do, any time you get distracted, is let that thought dissolve and come back to the breath. Ten times or a hundred times, it doesn't matter.

• Some thoughts seem a great deal more interesting than watching the breath. That's fine – there'll be plenty of time later for those thoughts. Just for now, return your attention to the breath. There's a subtle distinction between, on the one hand, a thought arising and, on the other, actively thinking about it – engaging with it. The same thing goes for sensations, sights or sounds. The sound of a dog barking or a pain in your lower back, for instance, can be a distraction if our attention jumps to it. But if we're able to just let it arise and pass in the background, then it doesn't have to disturb us.

• Spend some time observing your breath; practise acknowledging when you have become distracted and, then, practise gently guiding the attention back.

sound might take your attention away from the text message momentarily, but if your mind seizes upon it, in the next moment you might be whisked off into a memory of when some kids broke a window at the front of your house, then to when you hit a ball into a window next door when you were young. Some people think that they need utter silence and complete comfort to meditate well. This is another misconception. The mind can produce plenty of distractions even under the most 'ideal' of circumstances.

When it comes to distractions, it's not about how many there are or how many times you get pulled into one, it's about how you relate to them. Sounds are fine, thoughts are okay, wandering off is absolutely normal – as long as you just notice and quietly return the attention to your anchor point (the breath, in this chapter's exercise). If any judgements arise, try putting them to one side as well, and return to the direct experience of here and now.

How to Plan the Future in the Present

People often worry that if the aim in mindfulness is to keep our attention in the present, how can we reflect on what's happened or plan something in the future? It's a good question. However, it's totally possible to plan or reflect with your mind very much in the present. Let me give an example to illustrate mindful and unmindful planning. You're preparing for a stargazing holiday in a few months' time and you're on the computer looking up tour operators. You see a good package deal pop up on the screen that includes flights, hotel and a trip to the night-time desert. Instantly, your mind takes you there: you're in your hotel room getting everything ready – binoculars, star maps, woolly hat; and then you see yourself in the pitch-black night with a group of excited stargazers. The sky is dark down to the horizon and the stars are twinkling brightly. This is planning with your mind in the future.

Now, here's a more mindful approach: you're on the computer looking up tour operators and you see some deals popping up. Your mind jumps to an imagined view of the dark desert sky. You notice the fantasy, let it go and come back to sitting on your chair in front of the computer. You know you'd like to stay in a good hotel and spend a night in the desert with some like-minded stargazers, so you start to refine your search. Each time you find yourself fantasizing, you notice this and come back to the moment. You still feel excited by the prospect of future adventures, but avoid getting lost in imagined fantasies, which can easily solidify into expectations and which have the adverse effect of making you miss the lived experience of here and now.

Getting stuck in memories can be much harder to deal with than being swept off into fantasies of the future. We all have those intense memories that pop up, day after day or, perhaps, year after year. Typically, they revolve around embarrassing or regretful experiences or painful emotions.

Perhaps we once yelled horribly at a friend and keep coming back to the memory of the pain on their face. The emotion and physical response to the memory most often comes with a narrative revolving around questions such as 'why was I so stupid?' or 'what was I thinking?'. Often the memory involves (or has involved) a great deal of self-judgement and criticism.

For me, one of the memories that has consistently reared up over the years is that of a date I had at university. I'd grown very fond of this person and, one day, she invited me to her place for dinner. At the end of the evening I suggested that I head off, and she kindly offered to walk me to the bus stop. I think I gave her a peck on the cheek as I stepped onto the bus. Afterwards, the chemistry between us fizzled out and, despite my later attempts to keep it alive, it died completely. I was in psychotherapy at the time and I remember this date became quite a big theme in our discussions. I questioned again and again why it didn't go anywhere and what I did wrong. I came to the conclusion that a major factor was the distance I kept from everyone around me. I was able to relate well enough to her on a friendship level, but just not able to let it go any further.

For many years I deeply regretted how I'd acted – especially near the end of the night. I came to see that my stiffness and formality must have come across as a sign of disinterest, and that's why things fizzled out after the date. Of course, the reality may have been very different, but that's the memory

I'd built up. Because I berated myself for so long afterwards, the memory of that evening became imbued with all those years of self-judgement, making it even more emotionally intense. These days, when I manage to recall these memories through compassionate eyes, I recognize that I acted the way I did because my subconscious was trying to protect me. I didn't understand my behaviour at the time, but I was just doing my best. In the end, it was an event that fuelled my journey of self-inquiry, and through which I learnt a great deal about my habits and perceptions.

Forgiveness

Forgiveness is one of those tricky concepts that's easy to understand on the surface, but difficult and confusing to do in reality. As I see it, forgiveness is built into the practice of mindfulness: as we notice, acknowledge and let things be, we let things go. Forgiveness is an act of letting go. A few years back I felt I had reached the point where I was ready to forgive my father for his abuse when I was a child. I still know what he did was deeply wrong and very hurtful, but I realized that the anger I harboured was only hurting me. We arranged a meeting and I told him I forgave him for sexually abusing me. I'm not sure what he made of it, but it felt like a real lightening of some of the load I'd been carrying – a point of transition in my personal development. I can say from my own experience that sexual abuse isn't intransigent. As the great thirteenth-century poet Rumi said: 'The wound is the place where the light enters you.' With work, healing is possible and the learning you gain from dealing with the abuse can become a source of great strength.

If a memory surfaces, acknowledge it, together with the associated emotions and bodily sensations, but try not to feed it by getting sucked back into the details. Watch it through the eyes of kindness and acceptance as it arises and slowly dissolves back into your subconscious. This, to me, is an act of forgiveness.

Our 'baggage' (by which I mean all our hang-ups, painful memories, destructive habits, and so on) can't be dealt with in the abstract. Thinking along the lines of, 'when that memory arises then I'll . . .' just doesn't work. The abstract world is in the realm of thought, whereas our baggage is in the realm of embodied and lived experience – which only makes sense in the here and now. It's only in the present moment, as we inhabit our sensations and physical responses, that we can look at our 'stuff' with any authenticity. There is an old saying, 'Yesterday is history. Tomorrow is a mystery, but today is a gift. That's why it's called the present.' This teaching was beautifully presented by Master Oogway in the film *Kung Fu Panda*.

So, if the past no longer exists and the future is yet to come, what is this present moment we talk about? And what does the flow of time actually mean? To answer this, we're first going to look at what physics has to say about time.

What is Time?

In the seventeenth century, Sir Isaac Newton envisaged time as absolute across the entire Universe. Then, in the early twentieth century, Einstein published two theories that proved time is flexible and localised. Now, according to recent advances in physics, the very notion of time may just arise from our limited human perception.

The English word 'time' derives from an Indo-European root, meaning 'to divide'. People throughout history have divided time into units based on the seasons and the cycle of night and day, and our common understanding is that these divisions remain constant. But the idea of a steady flow of time with fixed units only solidified in the seventeenth century, after the work of the physicist Isaac Newton. He conceived of the divisions of time (and also of space) as absolute, rigid and immutable across the entire Universe, and inherent in his equations was the idea of time flowing from the past to the present. These Newtonian concepts were so influential that they soaked into our culture in the intervening centuries, such that they are now part of our common worldview.

But how does this Newtonian view square with our own experience? On a personal level, time can appear very elastic. The twentieth-century psychologist Mihaly Csikszentmihalyi coined the term 'flow' to describe a state of being where we're 'completely involved in an activity for its own sake'. He said that during this state, 'The ego falls away. Time flies. Every action, movement, and thought follows inevitably from the previous one'. Time does feel like it shrinks and expands around our immediate concerns, zipping by in some circumstances or dragging on in others. So it's clear our perception of time differs to Newton's fixed and immutable concept.

Time Is Flexible

It turns out modern physics also sees time very differently from Newton. The twentieth century witnessed developments and discoveries that have changed our understanding of all the characteristics that define what we thought of as 'time'. In 1905, Albert Einstein published his Theory of Special Relativity, which related the rate of the flow of time to speed. The theory shows that the faster you go, the slower time travels in comparison to a person at rest. Someone travelling very fast would therefore age slower than a person travelling slowly. In 1916 he went on to publish his Theory of General Relativity, which explained that time passes more slowly in a larger gravitational field, even for instances of the same speed. The theory predicts that time would pass faster for someone living high in the mountains than it does for someone living at sea level because mass (the mass of Earth in this case) distorts space and time. Although the effect is imperceptible to humans – the difference here would be expressed in billionths of a second – it has now been measured using the most precise clocks.

To understand this, imagine a blanket held out by its four corners. Now put a football on the blanket and see how it sits low in the taut material. Now add a tennis ball and watch how it rolls down towards the heavier football. If the football is the Earth and the tennis ball is you, then see how the mass of the Earth gently pulls you towards the planet's surface. We call this effect 'gravity'. Einstein showed that space and time are, in fact, two sides of the same coin. Just as the blanket (representing space) is stretched by the football, so is time. The greater the gravitational field, the more stretch it exerts and the slower time appears to pass relative to other parts of the Universe. When we stand at sea level we are closer to the centre of the Earth and experience more of its pull, so time travels more slowly. The most extreme example of this is around a black hole. A black hole is so huge that at its edge (a boundary called the event horizon) time appears to come to a complete standstill. For example, if an object were to fall into a black hole, from a distance it would seem as if it had come to a halt.

Newton thought there was an absolute time underlying everything and permeating the whole Universe, but Einstein proved that time and space are relative and flexible. There is no common time between different places (because gravity varies from place to place) and there's also no single time for any particular place (because the speed of that place can vary, relative to the rest of the Universe). The concept of 'now' cannot exist simultaneously across the Universe.

In our personal experience, time doesn't always seem to flow at a constant speed either. Sometimes it flies by and at other times it can feel like it drags on and on. Of course, this has nothing to do with Relativity; it's about the interplay between our attention, our interest and our ability in relation to what we're doing. In meditation, when we focus our attention on the present moment for an extended period, it can sometimes feel like time disappears altogether. Thus, time seems intimately connected with perception. To explore this, let's delve a little deeper into what time really is.

Entropy Is Time

In 1865, the German physicist Rudolf Clausius put forward the concept of entropy. Entropy is a measure of disorder in a system. Nature is designed in such a way that the entropy of an isolated system always increases over time; this is the second law of thermodynamics. Isolated systems always spontaneously evolve towards a state of maximum entropy. For example, an ice cube in a glass of water surrounded by room-temperature air is a relatively ordered system, and so it is 'low entropy' (ice is a crystal and more ordered than liquid water). Over time, the ice cube 'system' will naturally evolve towards a disordered state: the ice will melt and the water will gradually reach the same temperature as the surrounding air. Entropy, therefore, measures time; moving from past to future the Universe becomes, overall, less ordered. So, the question arises, why was the Universe more ordered in the past? It's the same question as: why does time travel forwards?

In his recent book *The Order of Time*, the theoretical physicist Carlo Rovelli puts forward the idea that it's all down to our perception: how ordered a system is depends on how you look at it. Imagine, for example, a set of address cards sorted alphabetically by last name. This will appear ordered to one person, but to someone who's looking for order by address location this will appear disordered. In a similar way, entropy exists because our perception of the world is only partial and we only see and interact with the world in a limited way. We look for 'order', expecting to find one particular configuration that conforms to our way of perceiving the world (the address cards ordered by last name), whilst remaining ignorant of the infinite other ways that 'order' could manifest (cards ordered by first name, in reverse alphabetical order, or by address). As Rovelli puts it, this is like having blurred vision – we're not seeing the full picture. Entropy is an expression of our ignorance of the world – what we see on one level is an imprecise blending of the myriad possible states, types of states and configurations on other levels and at other scales. Our perception of time and our notions of past and future, are themselves the result of blurred, incomplete understandings.

As an analogy let's consider the daily rise and fall of the Sun, Moon and stars. We now know it's not the Sun that moves – it's us. The movement is, in fact, an effect of perspective caused by the rotation of our planet. But it took many generations of study – and some of the pioneers who pursued the truth had to endure torture and ridicule – before it was accepted as fact. In fact, it still tricks us when we look up. But we have come to realize that appearances betray our incomplete view. Similarly, our perception of time might just be down to our incomplete view of the Universe. It may not be that the Universe, in general, is moving from low entropy to high entropy. It may just be that, in the particular subset of the Universe to which we belong (in which the address cards are ordered by last name) entropy is indeed increasing. Our myopic vision arises because we observe the Universe from within – we are of it; we can't see the wood for the trees. The flow of time is real, but only as real as the image formed by a pointillist painting or the dots of light on a computer screen.

BEING TIME

Is that the end of the story? Once again, it comes down to perception. From childhood, our entire mindset is steeped in past, present and future. But what if there's another way of seeing the world?

THIS THEORETICAL UNDERSTANDING of time further illuminates how limited our rational, everyday perception of things is. We perceive time passing from the past to the future; we see birth followed by death and existence followed by non-existence. This is what's known as a dualistic perspective, and characterizes our regular, ordinary mindset. With this view we think, 'That was then and this is now', 'This is me and that is you', 'That's a wall and that's a table'. It is characterized by discrimination and logic: *this* happened yesterday and will cause *that* tomorrow. There's nothing wrong with this world view, but as we've explored in previous chapters, it's only one perspective. There's another way of seeing the world – a non-dualistic view. In 1231, the Zen master Eihei Dogen wrote a text titled *Bendowa,* which is usually translated as 'The Wholehearted Way'. He said that, when someone sits in meditation (*zazen*) and is able to let go of their dualistic viewpoint and truly enter non-dual experience, 'he becomes, imperceptibly, one with each and all of the myriad things, and permeates completely all time.'

What does this mean, to 'permeate completely all time'?

Dogen taught that time does not exist separately from existence; things in the world do not exist *in* time, but *are* time. This is because things aren't really 'things', existing as entities separate from other things. 'Things' are actually processes – dynamic, interacting and always part of the whole. Things *are* time; the Universe *is* time. As an astronomer, people often asked me, 'What came before the Big Bang?' Because time only exists in the Universe, there is no time without the Universe – the Universe is time itself. So, the concept of 'before' the Universe is flawed. It's a nonsensical question. Similarly, asking 'What's outside the Universe?' is equally flawed. Space is inherent to our Universe, so there is no outside.

Dogen was actually rephrasing an aspect of Buddhist thought that dates back to the time of the Buddha, which says time as a separate phenomenon doesn't have a real existence. Twentieth-century philosopher Henri Bergson argued that 'duration' existed merely in our consciousness. This implies that the appearance of time flowing is just a trick of our memory and consciousness. There is only one time, and that is now. When the analytic, duality-creating mind quietens down, we can simply be in this moment without any discriminating concepts clouding our view. The seventh-century Zen master Hui-neng said, 'When both becoming and cessation cease to operate, the bliss of perfect rest and cessation of changes arises.' In this pure awareness, things and time are no longer separate, and the illusion of time flowing collapses. Everything in reality is to be found in the absolute now.

To see a World in a Grain of Sand
And a Heaven in a Wild Flower
Hold Infinity in the palm of your hand
And Eternity in an hour

WILLIAM BLAKE (1757–1827), POET
'AUGURIES OF INNOCENCE' (CA. 1805)

How does this relate to our practice of mindfulness? To begin with, the untrained mind is unfocussed and wanders off frequently. One moment we're drawn into a thought and, the next, whisked off into a memory or fantasy, causing us to miss the moment completely. We must apply effort to develop focus and concentration to the point where we can maintain a measure of steadiness. With this stable foundation, we can start to examine all the phenomena that arise within our awareness – bodily sensations, sounds, thoughts, memories, plans – without getting pulled off course. Out of that stability arises an awareness that is (somewhat) resistant to being sucked into the subject of our attention. We start to see that all these phenomena arise here and now, in this moment. Actually, with training, we see that even our distractions are part of this present moment. Instead of disturbing our awareness, they simply become part of the flow of arising and passing phenomena.

In the scientific (dualistic) viewpoint, a telescope is like a time machine. The light from the Sun takes eight minutes to arrive here; from the nearest star, it takes four years; and from the furthest reaches of the Universe, 13 billion years. On one level, when we look up we are looking into the past. But those light rays that have been travelling through the Universe know nothing of their history. They arrive now and deliver their information now. All of history, past and future, exists only now.

CHAPTER FIVE

Creating Reality

Mindfulness is the practice of seeing things clearly, just as they are. But what is it to see things 'just as they are'? Is that even possible? Isn't my view of the world always going to be different to yours? Is there one true Reality, or just millions of individual realities? In answering these questions, we must first explore how our senses work, how they are interpreted by the brain and how our perceptions can become coloured by a myriad biases, prejudices and delusions. Mindfulness gives us ways of seeing our delusions, letting go of our expectations and acknowledging our prejudices, so that we can touch the Universe as directly as possible, and marvel at its undistorted beauty.

Understanding our Senses

We receive information about the world from our senses. However, what arrives at our perceptual front doors is anything but faultless and complete. Our senses are certainly marvellous and very sensitive tools, but it's important to realize that there's more to the Universe than meets the eye (or ear or nose).

To give us a foundation to work from, let me introduce you to the Indian Buddhist philosophy of Yogachara. Yogachara developed in the fourth century, as people were trying to formulate ways of understanding the Buddha's teaching around delusion and enlightenment. Yogachara makes a useful distinction between the sense organ, the kind of information it collects about the world, and the 'consciousness' that arises when we start paying attention to that information. It describes eight different kinds of sense 'consciousnesses' that arise when our sense organs make contact with their different sense 'objects': (1) eyes (sense organ), sight (sense object) and seeing (sense consciousness); (2) ears, sound and hearing; (3) nose, smell and smelling; (4) mouth, taste and tasting; (5) skin, sensation and touching; (6) mind, thoughts and understanding; then there's (7) the concept of the self and self-reference; and finally (8) what is called the 'seed storehouse' (these last two will become clear as we go). When one of our *sense organs* (say our eyes) registers

an *object* (for example, the sight of the red giant star, Betelgeuse) we get the arising of 'sight consciousness'. When we pay attention to it, it's then interpreted by our 'mind consciousness', with reference to the seventh consciousness: our sense of self and our place in the world.

According to Yogachara, the normal way in which we see reality is only created when our mind touches what we sense from our environment and interprets it. Every individual's everyday interpretation, however, is necessarily influenced by past experiences, habits and desires. For example, we may either say, 'Wow, Betelgeuse is so red!' or 'That's boring'. Yogachara says that these perceptual colourations arise from our unconscious, or what in the philosophy is called the 'seed storehouse' (the eighth consciousness). This is vast and contains all our past experiences, tendencies, cultural conditioning, habits, likes and dislikes. In keeping with Buddhist teachings, the philosophy goes on to assert that it's possible to transcend these conditionings and colourations through practice and awakening to the truth of enlightenment.

For those of us that have yet to reach those lofty heights, however, as 'perceivers of reality' we are limited by the capacity of our senses to detect this reality, and by our ability to interpret and make sense of these impressions. As we get to know our reality more and more through our mindfulness practice, it's important to have a sense of what these limitations are and how they arise. This will allow us to under-

stand ourselves better and will give us the opportunity to see beyond them into the true limitless beauty of the cosmos.

There's More to the World Than Meets the Eye

Human sight is very specialized. Our eyes are sensitive to light between wavelengths of about 380 to 800 nanometres, because most of the Sun's energy arrives at the Earth's surface at these wavelengths. However, this visible range is but a miniscule fraction of the entire electromagnetic spectrum. Light with wavelengths below 380 nanometres is called ultraviolet (UV) and is easily seen by some insects, animals and fish. Bees, for example, see distinct patterns on flowers in the UV range, making them look very enticing; some species of butterfly that appear drab to human eyes look very attractive in the UV-sensitive vision of potential mates; and reindeer rely on UV light to spot edible lichens in the snow. If our eyes saw in the UV range instead of our current visible range, the world in general would be pretty dark, as the atmosphere is a very good UV absorber (which is good because too much UV radiation causes cancer, and that's why sunscreen is worth using). Beyond UV, there are X-rays and gamma-rays, going down to wavelengths of 1/1,000th of a nanometre.

Light with wavelengths above 800 nanometres we call infrared (IR); this was first discovered in 1800 by William Herschel. Being inquisitive, Herschel was experimenting with sunlight, shining it through a prism to make a spectrum

and then measuring the temperature of the light at different colours. He noticed his thermometer measured the highest temperature beyond the red end, where there was no visible light. This is what we now call the thermal-infrared, and although our eyes aren't sensitive to it, our skin is. The warm sensation that we feel on our cheeks in the sunlight or from a hot pavement in summer is IR radiation. IR cameras are tuned to see the heat energy emitted by people – airports increasingly use these to scan people as they arrive to see if they have a high body temperature indicating an infection. Beyond IR, there are microwaves and radio waves going up to wavelengths of thousands of kilometres. If we could see in microwaves we'd be totally blinded by the background 'noise' of TV and mobile phone signals that we are immersed in day-in and day-out.

If we could see in microwaves we'd be blinded by the 'noise' of TV signals

Our ability to hear the world around us is also constrained by the biology of our ears. Have you ever been out in the garden at dusk and heard the faint cheeps of bats flying around? Or the faint, deep rumble of a thunderclap in the distance? These are sounds at the very edges of our audible range. A healthy young person can hear sounds from frequencies of about 20 to 20,000 hertz (a measure of waves per second), but this sensitivity falls with age, particularly at the

high-frequency end. Bats emit sounds between 9,000 and 200,000 hertz, so out in the garden we are hearing only a tiny fraction of their chatter. And the same goes for dolphins, mice, moths and some birds. Most types of electrical equipment emit some kind of ultrasonic sound well above the human hearing range and, of course, ultrasonic scanners do just that. At the other end of the spectrum, some animals – such as whales, elephants and giraffes – vocalize using sound below 20 hertz. Volcanoes, earthquakes and ocean waves also generate plenty of sound that is too low for us to hear. What our ears are sensitive to, therefore, is only a small range of the whole audio spectrum out there.

Sensing the World as if for the First Time

It's important to appreciate there's more out there than what we can detect with our human senses. But these senses are highly tuned to the regions they can detect (barring my own nose, which really can't detect anything at all). This is what makes the tremendous sliding scale of sunset colours so dramatic, or the intricate harmonies of a symphony orchestra so compelling, or the skilful touch of a masseuse so delightful. When we open ourselves to sensing the world as if for the very first time (see the exercise 'Beginner's Mind'), then we may find completely new and wonderful ways of experiencing what might otherwise appear mundane.

MINDFULNESS EXERCISE

BEGINNER'S MIND

- Become aware of your sense of hearing. Imagine you've been deaf from birth, and by some miraculous procedure, have just been given the ability to hear for the first time. How wonderous that would be! Try listening to the world around you as if for the first time. Take a moment.
- Do your best not to label what you hear, just listen to the soundscape. Some sounds are obvious and some are subtle, some come from far away and some from close by.
- Science tells us our hearing has a limited range. Yes, we can accept that truth, and we can appreciate the diversity and beauty of the soundscape that we can hear. Enjoy listening with no effort. Take as long as you like.
- Now, become aware of your sense of sight. Imagine that you've been blind from birth and have just been given the ability to see. Blink open your eyes. Look at the view in front of you as if you were seeing it for the very first time. There's no need to move your eyes or head, just look at what's in front of you. Move your attention around the book into the periphery of your vision. Do your best not to label what's there – just look.
- Appreciate that your eyes are only able to sense a very limited range of the whole electromagnetic spectrum. But within those bounds, notice the variety of colours, shapes, tones, and brightnesses. There's an exquisite beauty in the everyday-ness of what's in front of you. Enjoy looking and seeing. Take as long as you like.
- Now begin to notice your other senses: what can you smell right now? What can you taste? What are you touching? Notice how there's a difference between the sense object (for example, your chair), the sense organ (your skin) and sense consciousness (the feeling of sitting).
- Isn't the range of experiences we can have just here, right now, absolutely incredible?

Now to Interpret It All

It's the brain's job to put all our sensory information together and interpret it. However, no matter how accurate (or inaccurate) our senses might be, interpretation can become skewed. Through encouraging us to examine our present-moment reality, mindfulness brings our filters and biases to light so we can accept them for what they are.

Perception is very much influenced by our underlying desires, expectations and past experiences, together with our need to objectify, label and categorize these experiences (what, in Yogachara, are seen as part of the seed storehouse). Optical illusions clearly demonstrate that the brain doesn't always interpret sensory information correctly.

Many scientific studies have shown that our expectations and desires change our perception. For example, when we see a desirable object – like a glass of water – we may perceive it to be physically closer than it really is. Other research suggests that the appearance of things we see can be influenced by our memory of other things we've just seen (in one study, participants misperceived the direction of motion of dots on a screen if, while viewing them, they were trying to keep in memory the direction of a different set of moving dots). There are numerous studies of how expectations affect us from the world of food and drink: people rate the taste of Coke higher when it's drunk from a cup with the brand logo;

and if we think a bottle of wine is expensive then we'll find it tastes more complex and rounded.

Anticipation of Experience

When we're told something (especially by an expert) it seems that we start to anticipate what our experience will be, and, in turn, our expectations change our physiology in order to prepare ourselves for that expected future. This is broadly what's going on in the 'placebo effect'. It's a feedback loop. Another example is our experience of pain. Pain is well known to correlate with what one expects or anticipates rather than being an objective measurement of the injury. And just because we see or hear something, it doesn't mean we actually register it. Selective attention is a type of mental filtering in which we fail to notice things we see or hear when they just aren't that exciting to us (for instance, your partner's request for you to do the washing up) when we're focussing on something else, or when they don't line up with our world view. There's a famous experiment (Simons and Chabris, 1999) that involves watching a video of two teams of people throwing a ball to one another while keeping count of the number of passes made by just one of the teams. While this is going on a person in a gorilla suit walks through the background of the scene. Astonishingly, only about half of observers actually notice the gorilla because they are concentrating so hard on keeping count!

In his recent book *Solve for Happy*, Google's former chief business officer Mo Gawdat made the point that our happiness is highly dependent on our expectations. Being a physicist myself, I have a certain fondness for equations, so I appreciate the formula he came up with: *happiness = reality – expectations*. What it says is that the fewer expectations we have (of any type, good or bad), the happier we'll be. This might seem quite a radical statement, but it's worth testing it out for yourself. It'll need some careful attention and for you to be honest with yourself.

Expectations & Past Experiences

Another thing that's important to realize is that what we perceive can sometimes be influenced by recollections of past experiences – so much so that new sensory input is obscured. The example used frequently in Buddhist teachings involves a rope coiled on the ground that is misperceived as a snake (which may have happened fairly frequently in ancient India). Equipped with your past experience of snakes – knowing that they're often found coiled up – when you see a 'snake' on the road out of the corner of your eye, you instinctively jump back, thinking that it's real. Your belief in a real snake, on the strength of a fleeting glance, can be strong enough to trigger a fully fearful reaction. However, if you manage not to run away immediately, you'll see your mistake – it's just a rope – phew! The existence of the snake was an illusion.

Whether we have just started practising mindfulness or have been practising for years, we all have expectations as to what mindfulness might do for us. Do you remember what you thought might happen the first time you meditated? Prior to coming to study with me, many people over the years have had one or two previous experiences of meditation that, quite often, they found pleasant and calming. But after practising mindfulness over a sustained period of time, they were surprised that they didn't always feel that level of pleasure or comfort. In such a situation, it's easy to jump to the conclusion that you're doing something wrong, or you're just not very good at it. But the opposite might be true. Mindfulness isn't about making you feel good (or bad). It's about encouraging you to notice how things are without changing them or judging them. Have you noticed any other expectations or illusions when practising mindfulness?

Hurting & Letting Go

When I first came to mindfulness in my mid-20s, I hadn't realized just how much my perception of the world was blinkered and limited. Not only was I subject to the everyday biases and tendencies that all humans are affected by, but I had also developed some rather cockeyed world views and attitudes, due to my unconscious reactions to the traumas of my upbringing. Since then, it's been a long and slow process of discovering, unpicking, hurting and letting go, letting go

some more, and developing a great deal of self-compassion and determination. Along the way I couldn't help but feel incredulous at how my brain had found such thorough ways of repressing my emotions and distorting my perception!

Whatever you discover on your journey, these two characteristics – self-compassion and determination – are paramount. You may uncover views of yourself as unworthy of love or friendship, unconscious emotional strategies for avoiding pain that are hurting yourself or others, habits of holding or blocking, or delusions such as thinking all you need is for someone to peer through your armour and see you for who you 'really' are. Whatever you find, do your best to see these things fully. Self-compassion means to acknowledge your pain and suffering, put a caring metaphorical arm around yourself, and feel what you feel without wanting to fix or change it. It's not easy. Ultimately, it's this kind of acceptance or loving kindness that's the key to allowing things to shift. And to follow this through, you need determination. You need to find the energy to persist in your practice even when you feel terrible and just want to shut it all out.

The more we resist how things actually are, the more we suffer

There is a Buddhist saying that 'pain is inevitable, suffering is optional', which can also (happily) be rephrased in the form

of an equation: *suffering* = *pain* × *resistance*. Like with any multiplication, if any of the factors fall to zero then so does the product – if *resistance* falls to zero then so does *suffering*. The more we resist how things actually are, or want them to go 'our way' (which ultimately stems from seeing ourselves as separate), the more we suffer and the more skewed our perception of the world becomes.

Still More Coming to Light

With continued practice, more and more will come to light. For example, I'm still discovering subtle ways in which miserliness affects my behaviour. A stark reminder of this came a few months ago when my wife and I were on holiday, making our way through a steep and spectacular canyon. As I was driving my wife kept suggesting I change down gear as we approached the tight hairpin bends. Time and again, though, I kept on leaving it until we were in the bend to change. There was a strange cognitive dissonance: I consciously understood this was not safe or efficient driving practice, but somehow kept finding myself doing it. It was only after we stopped and were talking about it that I realized that changing down meant increasing the engine revs and to my unconscious mind that meant higher petrol consumption that would, therefore, cost us more. Once I saw this and understood that it was a remnant of my parsimonious habits linked to emotional holding, I stopped doing it (much to my wife's relief).

JUST HOW MUCH OF THE UNIVERSE CAN WE SEE?

With incomplete knowledge and with limited technology we have to infer what's going on in the Universe; accepting that our knowledge of ourselves is incomplete, we can use mindfulness practice to infer what's going on in the mind. Mindfulness brings the otherwise unseen workings of the mind into conscious awareness so that we can see our habits, impulses and delusions.

WHEN IT COMES TO DETECTING what's out there in the Universe, astronomers are very limited. We can't go out there and shine a light on something or prod it to see what it does (except for maybe on the Moon or Mars). We're reliant solely on what arrives here on Earth. Astronomy is mostly concerned with looking at things that emit light – like stars – or things that absorb light from other, brighter objects in the background, like gas clouds. Some areas of astronomy are concerned with measuring particles that reach here from space, like charged ions (some of which cause the aurorae at the northern and southern poles) – but, mainly, astronomy gathers its information about the Universe indirectly. For example, the gravitational effects of large galaxies can bend 'spacetime' sufficiently to act like giant lenses to the light that passes by. By measuring how the light is bent we can infer the mass and shape of the lensing galaxy(ies). A combination of decades of measurements and theoretical predictions suggest that the Universe is made of only 5 per cent ordinary matter and energy. The remainder is currently understood to comprise 27 per cent 'dark matter' and 68 per cent 'dark energy'. 95 per cent of the Universe is therefore 'dark', meaning that it is unknown or not explainable by current scientific theory. This means that everything we've ever seen, measured or touched is made of stuff that amounts to just 5 per cent of the Universe.

The presence of dark matter is inferred from measurements of the speed of stars and gas orbiting in a galaxy. This rotation speed should decrease with distance from the centre, but in most galaxies (including our own) it stays relatively constant with distance. This suggests the presence of extra (invisible) material in the outer parts of the galaxy. The existence of dark energy has also been inferred through measurements of the expansion rate of the Universe. Contrary to expectations, observations indicate that in the ancient past the Universe was expanding more slowly than it is today. This means the rate of expansion of the Universe is accelerating, and this must be driven by a mysterious, as yet unknown (dark) energy that nothing in our science can so far begin to explain.

Bringing the Unseen Mind Into Awareness

Dark matter and dark energy are, in a sense, analogous to our mind. The mind is an intangible collection of processes, only a few of which enter our awareness, and only ever partially. To an unmindful person, much of its inner workings (including

our instincts, drives, desires and expectations – all part of the 'seed storehouse') are invisible. They might only ever 'see' its effects and consequences through impulsive actions. The practice of mindfulness works to bring some of these otherwise unseen workings into conscious awareness. We start to see our habits and tendencies more clearly and see how they motivate particular actions. In this sense, mindfulness is like a 'technology' that can enhance our sensory faculties. We become like astronomers, using technology to augment our limited sensing options.

In the last few years, a number of technological breakthroughs have allowed astronomers to measure gravitational waves – a totally new kind of information that comes to us from space. Gravitational waves are 'ripples' in spacetime, thousands of times smaller than the nucleus of an atom. This opens a new set of eyes on the Universe and with it the potential for helping us understand what makes up dark matter and energy.

The Fundamental Illusion

When I first started to practise Zen, I learned about the fundamental illusion that we all hold about ourselves and the world: that we are separate. Infants cry, burp and poo, and are 100 per cent 'at one' with the world. In this sense, babies are like any other animal. Necessarily, though, as the child grows, a sense of 'I' develops – this entity is different

from parents and friends at school. As the years go by, this idea of the self solidifies and, eventually, the fully-fledged individual is very clear about what it means. The definition of an adult is someone who recognizes their own 'self', lives independently, and knows what they're responsible for and what they're not. As important as developing a sense of self is, it comes with a downside. If we're not careful, it solidifies completely and ends up being fixed and unchangeable. Someone with this world view might think, 'I am the way I am, I've always been like this, and I can't possibly conceive that I'll die' (even though, on a rational level, they know that's not true). Alongside this fixed view of the self, a person develops the perception of separateness: 'This is me and here is my "edge", and that is you. You are not me.' We might feel like a solid snooker ball rolling across the table of life, knocking into other people with a clack, sometimes with a bang.

Zen, however, points out that these fixed and separate world views represent just one side of how things are. The experience of awakening (or enlightenment) is the process of seeing and experiencing life as fluid, dynamic and constantly changing, knowing that we are transitory manifestations of the one Universe in this moment. As my Zen teacher says, 'Rather than being a thing in a world of things, we are a process in a world of processes.' Another way to understand these two viewpoints – the separate and the non-separate – is by imagining a mountain range enveloped in cloud with the

rocky peaks protruding through the top of the veil. These peaks seem distinct and pretty much permanent – like the concepts 'you' and 'I', or objects like the wall or table. But if we could look using a different wavelength – say, the radio – the clouds would be transparent and we'd see how the mountaintops are connected to one another via the valley floors. The mountains are all part of the one range – 'you' and 'I', the wall and table, are all connected and just different parts of the whole. And the mountains are not really permanent. They change, albeit very slowly, through the action of the wind and rain. I remember realizing the reality of this dynamic, unified perspective on a retreat back in 2010, not just intellectually but by directly experiencing it. You can see through this deep illusion of perception too. It's not too difficult. Simply keep looking at your reality with curiosity and inquisitiveness.

As you practise more and more, your awareness will expand so that what was hidden becomes more and more conscious. Without judgement, you will become aware of your filters, biases, and limitations so that you will come to know yourself, and your experience of things, more directly. When you think and act in terms of a 'self' that is separate and unchanging, then this self will need feeding and protecting. The solution is to remember that you are always changing and none of us are truly separate from the Universe. Gradually, you are moving from a small and limited human state to a vast and unlimited cosmic state.

MINDFULNESS EXERCISE

OPEN PRESENCE

I invite you to try one of the primary meditation practices of many Buddhist traditions. It's often put front and centre because it is so effective at getting us out of our blinkered, thought-based world and into as direct a contact with our moment-to-moment existence as is humanly possible.

- Firstly, settle your body in a comfortable sitting posture. Lower your gaze and keep your eyes softly open looking at this page (or, if you've read this already, you can let them close fully).
- First become aware of the sensations in your body. There's no need to try and influence whatever you find in any way, just notice: your face . . . your hands . . . your legs . . . and feet. Now become aware of your breathing. Again, just notice it as it is right now. Any time you get distracted, just bring your attention back.
- We're now going to explore letting go of any fixed focus for your attention. If your attention was previously like a narrow beam of light, let your attention now become like a floodlight. Let go of consciously directing your attention. Instead let it be like a giant mirror, simply reflecting (being aware of) anything that passes in front of it – sensations, thoughts, memories, sounds – anything at all. In this awareness, you might notice the breath or other bodily sensations, but you're not deliberately pointing your awareness towards them. Take a moment before reading on.
- As you notice things, try not to grasp onto them. Create a space in your mind in which absolutely anything at all can arise and be seen. All you need to do on a conscious level is maintain your willingness to be open, acknowledge whatever arises and try not to interfere with it, label it or want it to be any different. Allow life to unfold within and around you. Again, take some time with this before reading on.
- Notice how, as you do this, your sense of self as a separate, fixed 'thing' can gently soften. You're melting into the Universe. You're no longer a small and limited human, you're becoming the vast, boundless cosmos.

CHAPTER SIX

PASSION & AWE

Science is full of emotion. It drives curiosity, and a good result or successful experiment can prompt feelings of elation, connection and satisfaction. Even the simple act of looking up at the night sky can elicit an array of emotions. Being mindful of our shifting emotional landscape is key if we want to understand ourselves better. When we react to painful emotions by denying them, ignoring them or distracting ourselves, we observe this and turn towards every experience with awareness and non-judgement. With this attitude of self-compassion, we can learn to make friends with suffering.

MARVELLING AT NEBULAE

Looking through a telescope at something in the vast depths of space is a special experience. But our eyes aren't very sensitive to colour. Taking a long-exposure photograph allows us to see its colour and detail. Both ways of looking at the Universe – with the naked eye and with the aid of technology – can elicit surprising emotions.

THE RETINA IS THE EYE'S LIGHT SENSOR. It is composed of rod- and cone-shaped cells: the cone cells are responsible for detecting colours and function best in relatively bright light, whereas the rod cells work better in dim light but have very little colour sensitivity. So, in low light – on a dark night when we're out looking at the stars – our eyes aren't very good at distinguishing colours.

If you ever get the chance to look at the Orion Nebula through a telescope, you'll see what I mean. With even a small amateur telescope, you see the four stars of the central Trapezium Cluster embedded in a patch of wispy nebulosity which, with the naked eye, has a washed-out grey colour. This is because there's not enough light to stimulate the colour-sensitive cone cells in your eye. A long exposure photograph, however, will reveal that the cloud is coloured a beautiful soft blend of blues and reds.

When someone looks through a telescope at something like the Orion Nebula for the first time, there are, in my

experience, two types of response: there's the 'wow!' and there's the 'is that it?' response. The 'is that it?' crowd are very likely comparing what they're seeing to a memory of a colourful image taken by a bigger telescope with a long exposure – and feeling slightly disappointed. It makes me sad when someone says 'is that it?' because a response like this shows that they have stepped out of present-moment experience – of observing something so vast and so distant it's barely imaginable – into comparison. They are no longer seeing the world just as it is.

Collecting Colour

A long exposure through a telescope collects much more light than our eyes can register in an instant, and gives us access to more detail. Using filters in front of the lens allows us to isolate certain wavelengths and produce a dramatic colour image. Professional narrow-band astronomical filters are designed to pick out the light emitted from certain elements – for example red light from sulphur, green light from hydrogen and blue light from oxygen. If you were to take a monochromatic (black and white) image of the Orion Nebula using a sulphur filter, then another with a hydrogen filter and another with an oxygen filter, then you would have three images that you could assign, on the computer, to the red, green and blue channels of a new colour image, and you would end up with a vivid, multi-hued representation of the nebula.

Colour images of nebulae made like this can be stunningly beautiful (and are also scientifically useful to boot). The enigmatic splendour and variety of these spectacles are often what draw people to astronomy in the first place, myself included. They are works of art. I remember walking down the corridors of the Space Telescope Science Institute in Baltimore, USA (where the Hubble Space Telescope is controlled), seeing huge framed pictures of various nebulae and galaxies, and being gobsmacked at their intricate beauty.

Contemplating a Nebula

Since you may not have a telescope or a dark sky available to you right now, I'd like to invite you to find an image of a nebula to contemplate. Try searching for the Hubble Space Telescope's twenty-fifth anniversary image of Westerlund 2 – it's truly magnificent! Westerlund 2 is a compact young star cluster in the Milky Way with an estimated age of 1 or 2 million years. Its stars – which were forming as humanity first evolved from apes – are lighting up and slowly dispersing the surrounding gas from which they were created. The effect is a superlative vista of sparkling lights against a multicoloured, velvety, watercolour-like backdrop. Other marvels include the Eagle Nebula (M16), the Carina Nebula (NGC 3372), the Tarantula Nebula (30 Dor) and NGC 346. If you're looking at one of these on the computer, make it as big as you can on your screen.

As you feast your eyes on the colours, shapes and patterns, try and sense your emotional state. You might feel awe – a fascinating emotion where feelings of reverence and personal insignificance mingle with exhilaration and wonder – or any other mix of emotions that might be simmering in the background from your day or recent life events. If you can identify an emotion, how do you know you're feeling it? What are the signals or signatures of it? What you're feeling may be very subtle, so take your time. Try not to judge whatever you find to be appropriate or inappropriate, good or bad; simply allow it. And if you feel unmoved by the image and can't identify any other emotional states within you, that's also perfectly fine.

Being Mindful of Emotions

Mindfulness encourages us to explore our emotional landscape with curiosity and acceptance, discovering its ever-changing highs and lows. Of course, it's very natural to want more highs and fewer lows, but this tendency to grasp or reject can cause us great suffering. Gradually, we learn to turn towards every experience with equanimity and kindness.

Most of the time, when we look inside we actually find a mixture of various emotions and feelings at different intensities, fizzling, tumbling and whirring away – and that's completely normal.

Emotions can be fleeting or they can be enduring (which may or may not result from a build-up of individual emotional experiences). For example, if you accidentally touch a hot stove, you might experience the emotion of anger at being so stupid, followed by a complex set of emotions connected to stories about how you never learn or how unlucky you are all the time; or you might chuckle and have feelings of self-compassion and forgiveness. Though it is likely that you had no control over the event itself, you do have some control over what your mind does after the event. This is where mindfulness can really help.

The Physical Feeling of an Emotion

One of the primary arenas for applying your mindful attention is your body. When you look inside, the first thing you notice might be an overall emotional 'tone' – like feeling annoyed or agitated – rather than a specific singular emotion. This may comprise a blend of many pleasant and unpleasant feelings, mixed with memories, ideas, beliefs, habits and tendencies. Whatever it is, there will always be a physical dimension to what you're feeling, even if it's quite subtle. If you're feeling annoyed, the physical sensations you might feel include heat or restlessness in the fingers or legs, or a hot bubbling in the belly. In contrast, joy or happiness might cause feelings of physical lightness, involuntary smiling or excited 'buzzing' or tingling sensations.

Focussing on the physical aspect of your experience helps keep you grounded in the present moment because sensations can only arise now. If you let your attention move to the associated thoughts or storylines then it's very easy to get pulled out of the now and into memories of the past or ideas about the future.

Staying Present

When my mum died last year, I was by her side in the hospital room. As intense waves of grief crashed over me, I knew I needed to keep my focus on the physical sensations. I felt the tears rolling down my face, the cries coming out of my mouth and the winded feeling in the pit of my stomach. Over the following hours and days, I recognized my habitual tendencies to shut down and repress my extreme emotions, avoid feeling the uncomfortable sensations, and to play the role of the strong, stable one around my relatives. Focussing on the physical allowed me to stay in the present moment and in touch with my feelings without being overwhelmed by them. It also helped me avoid getting stuck in states of disbelief, 'what-if' fantasies, and worries about the next steps.

Feeling positive and negative emotions are part-and-parcel of being human. In themselves, they're not necessarily good or bad. Let's take a traditionally 'negative' emotion like anger: it can feel horrible to be angry, but a burst of anger can also feel very satisfying and, even, appropriate in the moment.

Equally, feeling joyous can be wonderful, but the foreboding prospect of that joy coming to an end can feel dreadful. What actually causes the problem is our tendency to 'grasp' or to become 'attached' to the meanings we give to experiences. When we think of an experience as 'pleasant' we hold onto it and are afraid or sad about its unavoidable end. When we think of an experience as 'unpleasant' we want to push it away and get rid of the unpleasantness. Often, we'll do anything to deny it, to distract ourselves or to simply ignore it. Our natural responses – to push away negative or unpleasant feelings and emotions, and to glue ourselves to pleasant ones – is at the root of most of our suffering.

In our mindfulness practice we therefore try to turn towards each and every experience with a gentle, non-judgemental intention, not trying to push it away or pull it towards us. This is the kindest thing you can do for yourself. It's an act of self-compassion. The word 'compassion' comes from the Latin roots 'com', meaning 'to be with', and 'passion', which is the old word for pain or suffering (as in 'the passion of Christ'). Compassion therefore means to 'be with' pain, difficulty and suffering. When you practise self-compassion you do your best to be with suffering and make room for it all, no matter how messy or uncomfortable. Out of this might arise an acknowledgment that you need help or support (perhaps from a friend or therapist). Although no one can deal with this stuff for you, support, encouragement and

MINDFULNESS EXERCISE

INVESTIGATING SUFFERING

- Bring your attention to your posture. Notice the shape and orientation of your body. We're going to scan the body for any sensations of pain or discomfort. Guide your attention through your head, shoulders, torso, arms and hands, legs and feet.
- As your attention moves through your body, notice any areas of discomfort or pain. If you find multiple areas, try to identify the strongest one. If there's nothing obvious right now, keep scanning. You might have made a habit of ignoring a particular pain or uncomfortable sensations in general; if so, it may take sustained focus to bring it to mind. Like the background sound of a clock ticking that our awareness has filtered out, some sensations only rise into the conscious mind after steady attention.
- If you can't find any discomforts at all, look for the most intense physical sensation you are experiencing, in this moment.
- Now, put your attention right in the middle of what you've identified. Try to have an attitude of curiosity and inquisitiveness. If you feel resistance to looking at it head-on, that's fine. Try looking at it out of the corner of your mind's eye, or at the periphery of your attention.
- If you notice any thoughts or stories associated with the sensation, try to refocus on the physicality of the discomfort. What is the sensation like? Is it sharp or broad? Is it steady or throbbing or dancing around? Do the sensations change as you look at them?
- What about the 'level' of the feeling? Sometimes a pain can seem to intensify when we focus on it, but other times it can disappear. How does it feel in the surrounding parts of the body, such as the muscles, joints or skin?
- Bring as much kindness and gentleness as you can to the experience. It's not easy, and the exercise might take a few goes before you start to get the hang of it. Just do your best. You are trying something new, something that has great potential.
- As the practice comes to end, take a nice deep breath.

communication can help you find the courage and skill to bring about profound change, which begins with acts (even small ones) of allowing, witnessing and self-compassion.

Passionate Science

People become scientists because they are captivated by the beauty and wonder of the world around them and want to understand how it works. What motivates the scientist is common to all humans: an intense curiosity about the natural world. If we let it, this curiosity can connect us to the deepest mysteries of the Universe.

Scientists are often seen as cold, lacking in emotion, motivated only by reason and logic. This couldn't be further from the truth. People often become scientists because they are enthralled by the wonder of what they see around them. They're enchanted by the beauty of nebulae or spiral galaxies; they look at a Romanesco cauliflower, an insect's compound eye, the DNA double helix or a diamond crystal lattice, and are spellbound by the exquisite forms of nature.

Science is a very human pursuit and, as such, it is driven by emotions and feelings as much as anything else. Perhaps it is the lack of a public platform for scientists to share their emotional connection to their research that contributes to the misconception that they are unemotional. Pivotal moments in science – whether private or public – have always caused

spontaneous emotional and, even, physical reactions. In my career, the rejection of a grant proposal could bring on a multi-day depression, whereas an amazing result could engender waves of giddiness. In 1807, the British chemist Humphry Davy performed an experiment in which he electrolyzed molten potash, thus releasing a terrifying spray of fiery droplets made of a new element he named potassium. Davy reportedly experienced such a thrill at the discovery that he danced around the room in delight. It's only when scientists come to write up their work for academic journals that have very strict standards and style guidelines that they revert to the formal, impersonal passive voice ('the experiment was undertaken and the results show that').

The Great Adventure

Despite the way it may seem, emotion and scientific methodology are not mutually exclusive. Instead of opposing each other, emotions and logic can work to complement and reinforce each other. The quest for meaningful scientific results is a powerful and empowering one. Furthermore, emotions like inspiration, humility and awe cross easily into the realms of the spiritual. The famous popularizer of astronomy, Carl Sagan, once said, 'Science is not only compatible with spirituality; it is a profound source of spirituality.' Scientific enquiry can be a deeply life-affirming and enriching pursuit. It can connect us with the Universe on a level beyond facts

and knowledge, reaching towards the deep reality of what is. And you don't have to be in a lab or observatory – that connection can happen anytime, anywhere, if you let it.

A year or two ago I was on a week-long Zen retreat in the UK. The first few days followed the usual pattern that I have come to know well: a period of 'decompression' where the momentum of everyday life gradually subsides; physical discomfort and pain; arising of mental 'stuff' (memories, pain, issues) either spontaneously or prompted by the topic of the retreat; application of effort to stay with this discomfort and let it be; and then, finally, a level of ease. On this retreat I reached the beginning of the ease stage whilst we still had three or four days to go. I continued my gentle application of effort to stay with the process and things continued to quieten. Gradually I felt the edges of my body become less and less distinct and my sense of 'self' as a separate entity dissipate. I viscerally felt myself melting into the Universe: I was not just *in* the Universe or *part* of the Universe; I *was* the Universe. In fact, it was a gradual revealing of the fact that 'I' was not, and never would be, anything but the whole Universe. I felt an incredible wondrousness of being, whether sitting, eating, walking in the garden or having a cup of tea. The world took on a luminescent feel and in a period of

My fingers were like cosmic seaweed wafting on the current of the Universe

MINDFULNESS EXERCISE

JUST BEING

- Take a moment to make sure you're feeling comfortable. Check in with your posture and make any adjustments if you need to.
- Spend a moment or two appreciating this amazing Universe as it unfolds, moment-to-moment, in front of you. Whether your experience is one of ease or one of pain, you needn't wish for it to be different. Do your best to let things be just as they are. Remember, this is an act of kindness and self-compassion.
- You don't need to do anything. Let yourself settle into 'just being', wherever you are, right now. How do you feel? What's the emotional tone in your body?
- Let the seemingly distinct edges of your body soften as if the focus knob is being turned so that the view becomes blurred. Let yourself melt. See yourself as just one facet of the giant, sparkling jewel that is the entire Universe.

movement practice, one afternoon, that included some arm movements, I felt like my hands were waving through the Universe. My fingers were like cosmic seaweed wafting on the current of the Universe, passing through stars, galaxies and nebulae. It was only afterwards, as I shifted out of the direct experience and into the usual thought-based world, that the memory of the feeling was strange, and I realized what a profoundly spiritual experience it was.

CHAPTER SEVEN

MELT INTO THE UNIVERSE

When physicists started looking into the nature of light, their experiments brought into question the centuries-old paradigm of certainty and objectivity that underpinned all of physics and astronomy. The result was an existential crisis that lasted decades. Remarkably, the path of discovery and realization charted by these physicists parallels the journey of mindfulness and spiritual enquiry. This path begins when you identify a problem – perhaps your own existential crisis. The journey may take you over rough ground and around uncomfortable truths. But the crisis can only find resolution when you realize you're not just a separate part of this Universe. You are it.

REALIZING THERE'S SOMETHING MISSING

At the beginning of the nineteenth century, no physicist could have imagined how deeply their field would be shaken by new results over the coming century. Equally, the path of mindfulness has the potential to dramatically change the world views of any who take it up seriously.

THERE IS A FAMOUS PARABLE in the Zen tradition of the journey of a seeker, often portrayed as a child, to find a lost bull from their herd. Though difficult at first, as the journey progresses and the child finds, and slowly tames, their bull, they experience increasing harmony as their view of the world is transformed. The story of physics since the nineteenth century and the path of mindfulness that we have been exploring have some surprising similarities. New discoveries constantly challenge old beliefs and bring the potential to dramatically change the ways in which we view the world.

For many people, their first encounter with mindfulness follows the identification of a problem: life is too stressful, too fast-paced or too painful; perhaps it's become overly dull and automatic, or there's a feeling that 'there must be more to life than this'. Whatever it is, the mindful beginner has acknowledged their dissatisfaction and decided to act. In the first part of the Zen parable, the child realizes their bull is missing and starts looking for its footprints. This might be where you are now. It is impossible to underestimate how profound this moment is. If you think you have a long way to go, it is worth appreciating that many never even begin the search or realize that there is a journey to be made at all. Stepping over the starting line is the biggest step of all, and perhaps you have just made it.

The development of quantum mechanics begins in 1801 with some very puzzling 'footprints' of its own. At this time the British physicist Thomas Young performed an experiment in which he illuminated two parallel slits in a metal plate, behind which was a screen. On the screen he observed parallel bright and dark bands. The bands of light, Young determined, must represent an interference pattern, much like you'd see if two raindrops fell together into a still puddle and the circular ripple-patterns they made were to collide. For 150 years prior to this, light had been assumed to be composed of tiny particles called corpuscles, but this new experiment showed that light couldn't be made of discrete particles, but must be made of waves moving through space. Unbeknown to Young, this finding set the scene for the biggest shake-up of physics since the work of Isaac Newton.

In the next 80 years or so, scientists successfully made the paradigm shift to this new 'wave theory' of light. 1862 was a milestone year, when the Scottish physicist James Clerk Maxwell published his wave equations that mathematically described how light travels through space as vibrating electric and magnetic fields. Today, we call these electromagnetic waves.

In 1887, however, the German physicist Heinrich Hertz discovered something that seemed completely at odds with what Young had found and Maxwell had described. He illuminated a solid metal surface with varying frequencies of light and found that light above a certain frequency (frequency being equivalent to energy) would cause an electric spark. Hertz called this the 'photoelectric effect'. In 1905, Albert Einstein came up with (and was later awarded a Nobel Prize for) an explanation for this effect. Max Planck had already put forward a radical hypothesis that the energy emanating from any atomic system cannot take an arbitrary value, but must be limited to specific levels that he called 'quanta'. 'Quantum mechanics' was born at that moment, setting the scene for a whole new era in physics. Building on Planck's findings, Einstein theorized that, if light itself was made of individual quanta (particles later called 'photons') and if the energy of these quanta were above a certain threshold, they would have enough kick to dislodge electrons from the surface of a metal plate, thereby generating a current – or the spark that Hertz saw.

A quandary had appeared. Young's slit experiment showed that light was a wave that produced interference patterns, while Hertz's photoelectric effect together with Einstein's explanation showed that light must be composed of particles. So, was light made of particles as originally thought, or was it really a wave? Since the vast majority of information we receive from the Universe comes to us in the form of

light, understanding this fundamental problem was critical to astronomers as well as physicists.

Searching for the Answer

Questions around this 'wave-particle duality' problem led to much confusion and head scratching within the scientific community in the early twentieth century. In standard physics, a discrete particle (say, a snooker ball) had, at every moment, an exact position and velocity that varied with time, according to Newton's laws. But this didn't apply in the strange subatomic world where something appeared to be a particle in one instance and a wave in another – a new paradigm was needed.

In 1923 the French physicist Louis de Broglie put forward a mathematical theory, derived from Einstein's work on light and photons, showing that all particles can have wave characteristics – and vice versa. In 1925, a group of German physicists led by Werner Heisenberg developed a new framework for describing these subatomic wave-particles. The only way they could make it work, however, was by abandoning the conventional concepts of position and velocity in favour of abstract numbers known as 'matrices'. At the same time, the Austrian physicist Erwin Schrödinger developed an alternative (but equivalent) description. Schrödinger's equation described the way a system changes with time, not in terms of definite positions and velocities, but in terms of what is known as the 'wave function'. This defines the location of a wave-particle, at a given place and a given time, only in terms of its probability.

Precision and certainty, the bedrock of physics for centuries, were being abandoned and replaced by probability. This was anathema to many physicists at the time, including Einstein who, in a letter to Max Born, wrote that God 'is not playing at dice' with the Universe. In 1927, Heisenberg did come up with a principle describing the point at which the conventional concepts he earlier abandoned (like position and velocity) could be applied in the subatomic world. This became known as his 'uncertainty principle'.

Inadequate Language

The ambiguity of the wave-particle concept – and the inherent uncertainty with which we measure it – arises (as Heisenberg himself understood) because we project our discriminative view of particles as one type of thing (a discrete object) and waves as a completely different thing (a disturbance moving through a medium) into the subatomic world. In the late 1920s, physics was showing us directly that this isn't the way things are; that, actually, quanta could be both a particle and a wave. All previously held notions and world views just didn't apply at this level, and the mathematical solutions that were found only made sense if conventional concepts, or definite measurements, were replaced with abstract maths and imaginary numbers and probabilities. It became obvious that language and symbols were themselves inadequate.

What physicists hadn't realized (and what we ourselves often don't realize) is that all language is inherently based in a dualistic world view. Sentences are constructed from subjects and objects, for example: 'I see that' – there is an observer (I) and the observed (that). We are innately discriminating between 'this's – 'I have this problem' – and 'that's – 'I'm looking for that solution "out there"'. In physics, the 'this' was a particle and the 'that' was a wave.

Fundamentally, language is a tool and its subject/object construction has proved extremely useful over the millennia to describe the world, make abstract thoughts and communicate them. However, it's important to realize that, built into language, is the mindset of separateness and distinction. We feel that we are separate to that which we see. It turns out the solution to the apparent problem, both in our spiritual quest and in the story of quantum mechanics, lies in realising that this worldview is just that – one amongst many ways of seeing things.

Following the Footprints

Let's say you're stressed out and a friend recommends mindfulness. Metaphorically, you've now identified the footprints of the lost bull and are following them. Practically, you begin by working on developing focus and concentration and applying patience and kindness to your experience. With continued practice, your attitude to life starts to soften as you begin seeing habits and tendencies with clarity and compassion. Gradually your relationship to all experiences and thoughts begins to shift, and you might start asking questions like: Why do I suffer so much? Who am I? Where do I stop and the world begin? These deeper questions provoke a kind of restlessness and a concomitant motivation to come face-to-face with what's arising.

Following the footprints with patience and determination will eventually lead you to the bull itself. At first, you might just catch a glimpse as it disappears around a distant rock, but if you persevere you'll eventually be able to find it and catch it. All bulls are different of course; some are docile but others can be real bucking broncos. In our physics story, this stage corresponds to the developments made by people like Schrödinger and Heisenberg who deeply challenged the long-held views of science being about distinction, certainty and repeatability. The questions that were being asked were: 'When does a particle stop being a particle and start being a wave?', 'What does it mean for something to be a particle *and* a wave?', and 'What does it mean for a 'thing' to have an indeterminate position?' You could say the bull of quantum mechanics was very difficult to capture and tame.

In our spiritual enquiry, the bull represents that which you look at when you ask the question, 'Who am I?' – which also has an answer that transcends the duality of language and mind-based thought. For some, the 'I' that they find may be reasonably quiescent, but for others it may be like a spitting bonfire or a raging sea, full of pain or regrets from the past or anxieties about the future.

Taming Our Metaphorical Bull

If you stick with your enquiry long enough, through thick and thin and with openness, compassion and patience, then eventually your experience of your mind will calm down. You'll have tamed your bull sufficiently to begin riding it home. By continuing to face this 'I', at some point the distinction between observer and observed will evaporate (perhaps when you're least expecting it) and you'll glimpse a new world of oneness or 'non-duality'. Stepping towards this direct embodied experience (which is wholly different from a purely intellectual or conceptual understanding) you see how there really are no separate things. We see the Universe as one ocean of dynamic, interconnected processes and that there is no distinct edge to 'me' – I am blended with this Universe.

Of course, science has encountered suggestions of this all along. For example, even in conventional physics, the force of gravity is understood to diminish with distance, but never completely to zero. This means that, however infinitesimally small it becomes, the gravitational pull caused by something even as small as your own body is felt throughout the entire

Universe. If your body disappeared, the entire Universe would be affected.

When you breathe out, at what point does the air you expel stop being 'you'? When you hear a sound, is that sound 'inside' or 'outside' of you? Is the gravity that holds you to the ground part of you, or is it separate? It might be that, as you melt into the Universe, these questions become as meaningless as asking whether a wave is separate from the ocean or not.

The direct experience of oneness is like sipping cold water and being (not just knowing) the cold. You know you *are* the Universe. In the parable, the child becomes more and more familiar and comfortable with the bull until, at some point, the child clambers on top of the bull and they become as one. The child ends up at home, sitting peacefully and content, immersed in the beautiful view.

It seems Heisenberg touched upon this kind of direct knowing himself. In his 1958 book *Physics and Philosophy*, written after he had visited India and encountered Indian philosophy, he wrote, 'Now we know that it is always the same matter, the same various chemical compounds that may belong to any object ... The world thus appears as a complicated tissue of events, in which connections of different kinds alternate or overlap or combine and thereby determine the texture of the whole.' On a fundamental level, Heisenberg's uncertainty principle is a measure of the extent of the interrelatedness of the Universe. This is what we find when we

wake up to the non-dual perspective: there are no 'things', only processes; the Universe is one vast, interconnected, interdependent web in constant motion. But finding this pure, non-conceptual perspective is just the next step on the journey. It's not the end.

Channel-hopping

We now have two equally valid perspectives: one saying 'I am me and different to you' (or equivalently: it's a particle *or* it's a wave), and the other that says 'I am the whole Universe and not separate from anything' (it's both a particle *and* a wave). The next step in our practice is to familiarize ourselves with switching channels from the conventional (dualistic), phenomenal world of relative truth to the non-separate (non-dual) viewpoint, and back again. It's similar to those optical illusions where you first see two white faces looking at one another, then a moment later a black wine goblet appears in the space between the faces. After the controversies of the 1920s and 1930s, the Danish physicist Niels Bohr concluded that everything has both wave-like and particle-like aspects – they are complementary features like two sides of one coin. However, if we remain channel-hopping between the dual and the non-dual viewpoints then we're essentially still stuck in the world of duality, of there being two viewpoints. There is more in the story of the child and the bull, just as there is more to the story of quantum mechanics. Let's explore.

How to Make Sense of it All

The nature of the Universe knows nothing of distinction, division or discrimination. It just is – whole and complete, one single system. It is in this direction that we must go in order to make sense of things, and it requires us to open our eyes to a new perspective.

You could say that ever since Thomas Young inaugurated quantum mechanics in 1801, physicists have been making increasing efforts to 'transcend' the apparent dual/ non-dual worldviews they have uncovered by coming up with ever-more subtle 'unified theories'.

According to the Schrödinger equation, prior to measurement, any wave-particle (an electron, say) has a probability of being found at any given location. These probabilities evolve with time as a combination of different possible states – meaning there's a probability of the electron being here *and* a probability of it being there. However, when anyone actually measures the system, it will always be found in a definite state (the electron is found here, *not* there). This constitutes what has become known as the 'measurement problem'. The act of measurement somehow causes the system to choose exactly one of all the many possibilities.

How this works is still hotly debated within the quantum physics world today. The standard interpretation dating back to the 1930s is that the interaction of an observer external to the quantum system is what causes the choice to be made (or, technically speaking, the wave function to collapse). But this begs the question of what constitutes an interaction or even an observer, and reminds us of the inherent semantic limitation of a language based in subject/object separateness and distinction. The American physicist Hugh Everett proposed an alternative theory in 1957 called the 'many-worlds interpretation'. This states that every time a measurement is made the wave function doesn't collapse, but the Universe 'splits' into as many different time-lines as there were possibilities. We continue in just one of these universes and hence measure only one result. As fascinating as this theory is, it doesn't resolve how the split actually happens.

Quantum cosmology is a more recent idea, put forward by theorists John Wheeler, Stephen Hawking and others towards the end of the twentieth century, which attempts to chart a more holistic path. Rather than considering isolated quantum systems, quantum cosmology views the entire cosmos as one quantum system. In this theory, the information, the observer and the observed, are inextricably part of this one Universe and therefore mutually interdependent. This bears some comparison with the Buddhist view. The Buddha taught that the subject, object and everything we can know about the Universe is 'empty'. 'Emptiness' is a Buddhist term meaning 'to be absent of any inherent or essential quality'. Subject and object aren't 'things', separate and distinct (like isolated

quantum systems and observers) but are ever-changing processes that arise only through their interdependence on everything else going on in the entire Universe. Emptiness is synonymous with the ever-changing state of flux that gives everything life and dynamism.

Relativity, Interconnectedness & Impermanence

In 1929, during his lecture tour of India, Heisenberg spent some time with the celebrated Indian poet Rabindranath Tagore. From his discussions with Tagore and upon later reflection, he recognized that the concepts of relativity, interconnectedness and impermanence – the very things that were proving so difficult for the physics community to grasp – are fundamental aspects of our physical reality. Not only that; they form the very basis of Indian spiritual traditions like Buddhism and Hinduism.

As the twentieth century progressed, the study and application of quantum mechanics exploded. In the 1940s, spearheaded by theoretical physicists like Paul Dirac and Richard Feynman, quantum mechanics was shown to explain other electromagnetic forces. This expanded theory subsequently became known as 'quantum electrodynamics' (QED). QED includes not just a quantum theory of light as an electromagnetic wave, but an understanding of electrons, electrical charge and magnetic fields, in agreement with Einstein's special relativity. In the late 1960s and 1970s, the theory was expanded to include the weak nuclear force, resulting in the so-called 'electroweak theory'. In the 1970s, physicists developed an explanation of the strong nuclear force in terms of quantum mechanics, which became known as 'quantum chromodynamics'. Of the five 'fundamental forces' – electricity; magnetism; the weak and strong nuclear forces; and gravity – we now understand four in terms of quantum mechanics. These advances have been critical in furthering our knowledge of the early evolution of the Universe. The search is now on to unify the electroweak theory and quantum chromodynamics, and then to come up with a quantum theory of gravity.

What these theories show is that the five fundamental forces, which, under normal circumstances, seem distinct, are actually deeply interconnected. Discrimination and reduction is useful in physics, but only to a certain extent. In reality the Universe isn't divided into distinct parts; it is one whole. Physics is reaching towards its own kind of oneness: a coherent theoretical framework that fully explains and links together all physical aspects of the Universe. This 'transcendence' parallels the next step in your spiritual journey: to see beyond the dual/non-dual distinction.

Going Beyond

The way to see both separateness and togetherness on a spiritual level has been known and practised for centuries. In the parable of the lost bull, the child and the bull merge and, having returned home, the traveller sits in repose, admiring the view. Realizing our oneness, our total integration with this Universe, can be tremendously beautiful and serene. Just like the child, we can feel like we've arrived home. But the child is still there – there's still a 'me' in the picture, existing, observing and feeling.

At some point, there comes a moment when this 'me' also disappears. The traditional Zen commentary says that the seeker, the view, the home and the serenity all merge into 'no thing'. This moment is often pictured as the famous empty circle of Zen. Nothing can be said of this because it's beyond language and labels, and beyond the concepts of 'separate' or 'not separate', dual or non-dual. All boundaries between inner self and external world have collapsed. The famous sixth-century Taoist philosopher, Lao-tzu, was speaking of this stage when he wrote in the *Tao Te Ching*, 'The Tao that can be told is not the eternal Tao; The name that can be named is not the eternal name.'

Emerging from emptiness (symbolized by the circle) back into the bright light of the world, the Zen commentary says we've reached 'the source' and 'dwelling in one's true abode' we see that 'the river flows tranquilly on and the flowers are red'. Nothing changes – the river still flows, the mortgage still needs paying and the laundry still needs doing – but our perception of it all is radically different. We can live with ease

MINDFULNESS EXERCISE

MELTING INTO THE UNIVERSE

- Settle your body into an upright, balanced and relaxed posture. If you've read this already you may want to let your eyes close fully.
- Start to direct your attention inwards. Soften your face, your shoulders and your belly. Take some time to let yourself arrive. Notice the weight of your body pressing on your seat, or on the floor. Check in with your breathing. If your mind seems busy with thoughts, don't worry. Keep returning your focus to your breath, knowing that each time you notice yourself wandering off is a moment of awareness.
- Now, let go of your focus on the breath and let your attention broaden out. Let go of any conscious directing of your attention. If attention on the breath is like walking rhythmically, step after step along a clifftop path, then letting go of this focus is like an eagle spreading its wings, stepping off the cliff and soaring out over the ocean. In this vast, open awareness, you can allow absolutely anything at all to arise, be seen, and dissolve when it's ready. The only thing the eagle needs to do to keep soaring is maintain its outstretched wings – all you need to do is maintain this openness with a willingness to acknowledge whatever arises.
- You may notice the breath and other sensations and feelings, but you're not consciously angling your attention towards them. You may notice thoughts, memories, ideas or plans; sounds, sights, twitches, or any manner of phenomena. Give them all space to be. Notice them coming and going. Any time you find yourself actively thinking about something, just let the thought-train dissolve and allow your attention, once more, to become broad and open.
- Fully immerse yourself in the experience of being – enjoy the natural unfolding of each moment before you. Allow yourself to effortlessly and fully melt into the Universe.
- As you come to the end of the exercise, gently become aware of the edges of your body and take a nice deep breath.

and lightness, knowing that on a fundamental level, there are no problems. The final step in the parable of the lost bull shows the seeker reborn and restored, mingling with others in the ordinary world, but with a new understanding of things.

Just because you're approaching the end of this book, it doesn't mean that you are coming to the end of your own bull-taming journey. I invite you to return to the mindful exercises and try them again. Perhaps spend a whole week practising one, then move onto the next. And remember, you're not separate from this vast, dynamic Universe, you *are* it – the water in your body is made from hydrogen that was created during the Big Bang and oxygen that was formed in the cores of long-dead stars.

Emptiness here, Emptiness there,
but the infinite universe
stands always before your eyes.
Infinitely large and infinitely small;
no difference, for definitions have vanished
and no boundaries are seen.

So too with Being and non-Being.
Don't waste time in doubts and arguments
that have nothing to do with this.

One thing, all things,
move among and intermingle without distinction.
To live in this realization
is to be without anxiety about non-perfection.
To live in this faith is the road to non-duality,
because the non-dual is one with the trusting mind.

Words!
The Way is beyond language,
for in it there is
no yesterday
no tomorrow
no today.

SENG-TS'AN (SIXTH CENTURY), ZEN MASTER
'HSIN-HSIN MING' ('FAITH IN MIND', TRANSLATED BY RICHARD B. CLARK)

BIBLIOGRAPHY

Solve For Happy by Mo Gawdat
(North Star Way, 2017)

The Second Brain by Michael D. Gershon
(HarperCollins, 1999)

Physics and Philosophy: The Revolution in Modern Science by Werner Heisenberg
(Penguin, London, 2000)

The Master and His Emissary: The Divided Brain and the Making of the Western World by Iain McGilchrist
(Yale University Press, 2009)

The Polyvagal Theory by Stephen Porges
(W. W. Norton & Co., 2011)

The Order of Time by Carlo Rovelli
(Allen Lane, 2018)

INDEX

Acknowledgements

I'd like to thank my wife, Jo, for her deep insights and many stimulating discussions around various topics covered in the book. I'd like to thank my author-friend Livi Michael for her guidance, encouragement and careful reading of earlier drafts. I'd also like to thank my editors for their thoughtful, thorough and insightful comments that greatly improved the presentation of this book. I am and will always be deeply grateful to my Zen teacher Julian Daizan Skinner for all his wisdom and guidance over the years.

Credits

'Faith in Mind' excerpt reprinted with permission of White Pine Press